Your Microscope Hobby

HOW TO MAKE MULTI-COLORED FILTERS:
RHEINBERG, POLARIZING, DARKFIELD
AND OBLIQUE

Michael W. Shaw

Fresh Squeezed Publishing
RICHMOND, VIRGINIA

Copyright © 2014 by Michael W. Shaw.

All rights reserved. No part of this publication may be reproduced, distributed or transmitted in any form or by any means, including photocopying, recording, or other electronic or mechanical methods, without the prior written permission of the publisher, except in the case of brief quotations embodied in critical reviews and certain other noncommercial uses permitted by copyright law. For permission requests, write to the publisher, addressed "Attention: Permissions Coordinator," at the address below. Author and publisher not responsible for any accidents or damage due to use or misuse of materials or tools described herein.

Michael Shaw/Fresh Squeezed Publishing
PO Box 742
Midlothian, VA 23113 USA
www.tardigrade.us

Ordering Information:

Quantity sales. Special discounts are available on quantity purchases by corporations, associations, and others. For details, contact the "Special Sales Department" at the address above.

Your Microscope Hobby/ Michael W. Shaw —1st ed.

ISBN 978-1511421478

Contents

Publisher's Note 3

Introduction 5

Benefits of Rheinberg Filters 11

Microscope Requirements 19

Tools Required 39

Filter Making Materials 63

The Filter Making Process 95

Polarization 137

Oblique or DIY/DIC 141

Cases 159

Selling Filters for Fun and Profit 171

How to Make a Microscope Camera Adapter 213

How to Make a Plant Press 237

Websites 243

Acknowledgements 249

To Donna

Publisher's Note

This print edition is in black and white. Since the subject of this book is primarily about adding color to microscope viewing and photography, the reader is at a disadvantage. To the buyer of this book the author and publisher must somehow make amends.

Rather than key-code the pictures using patterns of black hash lines and dots, we are making this book available in its full color electronic version for free to owners of this paperback. It would have been impossible to make a print version in full color for an affordable cost to the reader. We feel that this black and white version allows bench top and casual reading, while the color e-book version provides the necessary clarification regarding the use of colors.

To obtain the free full color version in the format you desire, check the instructions in the back of this book.

Introduction

What are Rheinberg Filters? Although you may already have some idea about them, I'll just say a few words to highlight what they do and how they work.

Microscopes use light, passing through mostly transparent objects. Practically all of the microorganisms in a pond, micro fossils called diatoms, pollen spores, crystals, and many of the other interesting specimens we want to look at, are clear. So these very beautiful things are very hard to see against a white background of pure light passing from the illuminator through your eyepiece. One way to create a big difference between your subject and the background, called contrast, is to use chemical staining. This is where you dip your specimen in some type of dark chemical, and it physically changes the color of the specimen. Another way to achieve contrast is to interfere with the light waves, and this is called *Phase Contrast*. There is another excellent way to block out some of the light, called *Darkfield*, using a special condenser. Then there is a way to aim the light at an angle, called *Oblique Lighting*. You can also interfere with the light beam using *Polarization*, which is how your sunglasses cut out glare, and how you see 3D movies in the theater. Yet another way is called *Differential Interference Contrast*, using a type of polarization too. This book is about *Optical Staining*. That's a fancy way of saying we color the light, selectively. It's like dyeing cloth.

This light staining is called *Rheinberg*, named after the man who created this technique in 1896, Julius Rheinberg. We selectively make the specimen one color, and we make the background a contrasting, darker color.

Staining, in the traditional sense using chemicals, is the best way to see certain things like blood cells and bacteria, and certainly is a great way to look at many microorganisms. There are three problems with traditional chemical staining. The first is that it is messy and difficult to do well. Next, the chemicals are dangerous and highly toxic. The third problem is that you have to kill anything you stain, and a dead specimen is just not as cute as a live one.

This is why optical staining is a fabulous method, and that's what Rheinberg filters do. They create the much needed contrast without the mess, and without the high priced equipment needed in the other above mentioned techniques. A Rheinberg filter is really two filters in one: an outer filter and an inner filter. The outer filter, which is called the *annulus*, determines the color of the specimen. The inner filter, called the *center stop*, determines the color of the background. Can you imagine what it would be like if you could make your specimen any color and your background any contrasting color? Well, don't imagine it, do it!

There are many great resources on the web and in books that describe Rheinberg filters, how they work, and how to make them. This book saves you the trouble of doing all that research, and reading. Throughout this book, and listed again in the reference section in the back, are some good resources if you are curious to learn even more. But this book was created to save you time and effort by giving you what you need to know right now. And you can trust me, because I am the expert on the subject.

For many years I've manufactured and sold sets of Rheinberg filters for the various makes and models of microscopes, new, old, and antique. When the famous author and microscopist Mort Abrams needed a set of Rheinberg filters for his extremely rare *Mikropolykromar* attachment, he came to me. Using extreme precision to meet Zeiss standards, I created a custom set of filters in colors he never imagined he'd have for a device designed almost a century ago. For the past ten years, I've made filters for universities around the world, for the Mayo Clinic, for professors, industrial scientists, hospital researchers, doctors and veterinarians on every continent, and for amateurs of course as well. In fact, I'm the only one in the world at present who regularly manufactures and sells Rheinberg filters to specification. This book not only will tell you how to make your own Rheinberg filters, but will allow you to make them for others as well, the same way I do. Here's a quote from one of my customers:

> *"Received the set last weekend and am very pleased! A reasonably priced quality product above and beyond my expectations. I will be using them for years. A great addition to my microscopy accessories. Thanks again!*
>
> *—J Wilhelm"*

Throughout this book (and again in the back) I list all my suppliers and resources. To make it easier, however, I've also listed many of the tools and raw materials in a special on-line microscope store to support this book. You will notice as you read that I often make reference to my Amazon store, when an item discussed. The basic store link is:

http://astore.amazon.com/mikesmicroscopestore-20

Don't feel obligated to buy from my store, because I want you to find the best deals and be creative, and that makes it so much more fun. I've created the store, however, so you can immediately see the product detail and the most current prices. To show prices in this book would take up too much space, and this book wouldn't be accurate when prices change. So I've created the store, which will always be current in description, availability, and price.

If you do want to purchase something from my store, you'll be buying through Amazon, though they pay me commission on certain items. That's in the spirit of full disclosure.

Whether you bought this book because you are simply curious and want to add to your knowledge, or because you want to make a set of filters for yourself, or because you wouldn't mind selling some filters, you've made the right decision. This book is essential for all of the above reasons. And remember—a set of my filters now sells at ten times the price of this book. As far as value is concerned, you are already ahead. There is a saying:

Your Microscope Hobby

> "No one ever really paid the price of a book—only the price of printing it.
>
> —Louis I. Kahn"

Here is one more quote to get you inspired:

> "The filters just arrived and I have to say that I am ABSOLUTELY FLOORED and THRILLED at the quality and quantity of the product. I am completely blown away. You outdid yourself and I will recommend you to anyone and everyone I can. THANK YOU! Fantastic work! THANK YOU AGAIN!
>
> —Robb"

Okay. Let's get started. As you go through the material, think about your own geography—where you live in relation to suppliers in your area. Think about your microscope, and how you are going to test your filters. Consider ordering from the companies that I recommend. I receive no commissions or special discounts from any company I mention, but I do get a commission from my Amazon store which has no effect on your price. I'm simply sharing with you all my sources.

Why am I giving you all of my trade secrets that took me years to develop? Because I can't go on making filters forever, and I'd like to think that passing on my knowledge will help some folks to take up where I leave off. The world needs Rheinberg filters.

CHAPTER 1

Benefits of Rheinberg Filters

Whether you are a working professional or an avid enthusiast, you can use Rheinberg filters in your research. By learning some of these filter making techniques, you will always be able to create filters of high quality. You'll be able to do polarized light microscopy as well, because all you need is a sheet of polarizing material.

You will begin to look at the world in a new way, because when you go into a chain drug store, an office supply store, or a crafts store, you'll be looking for materials that you can use to make filters. Even a trip to the hardware store to buy lumber will land you in an aisle you never went to before, because you'll be looking for new tools to make filters. Those CD's that you used to throw away can now be recycled for filter making, their clam-shell cases serving you as polarization interference filters. Those little Styrofoam circle plugs at the bottom of the CD towers will be valued spacers for your filters.

In essence you'll see the world through Rheinberg filters (which are almost as good as rose colored glasses)!

Perhaps you would like to enter some of your photomicrography in a contest, or even sell some of the photos you've taken through the microscope. Now you are on the fast track, because Rheinberg filters will give you the *pzazzz* that you need. There are two great contests that you should enter each year, no matter what your level in the field of microscopy.

    Nikon's Small World Contest at:
    http://www.nikonsmallworld.com/

    Olympus' Bioscapes Contest at:
    http://www.olympusbioscapes.com/

The Nikon contest sends you a beautiful wall calendar with the winning photos, simply for entering the contest. The Olympus contest also offers different giveaways just for entering. Certainly the prizes and runner up prizes for these contests are a dream come true. Both contests give away thousands of dollars in cameras and equipment.

One thing you will notice about the contest winners is that many of them are amateurs. If you subscribe to the microscope forums, or are active in the world of microscopy, you will see your friends and acquaintances among the winners and honorable mentions. Why not you?

Do you have a website, social network page, or blog? Certainly, these are inexpensive or free in some cases, and this is a great way to post some beautiful Rheinberg pictures. Guess what? I have had many emails from people interested in my photos, for publications, science projects, and best of all—buyers.

I actually sold a collection of my Epsom Salt crystal photographs to a restaurant in England. A touch of irony is that this restaurant is located in the town of Epsom, England (where the salt originally came from). You can generate interest in your photos simply by posting them on your website. If you use some key words and show your email address, you could get inquiries from all over the world.

If you are ever in Epsom, Surrey, England, be sure to stop by Pizza Express. Besides the great food, they have some really nice décor—namely my photos. Mention my name and you'll get a good seat.

http://www.tripadvisor.com/Restaurant_Review-g504170-d732740-Reviews-Pizza_Express_Epsom-Epsom_Surrey_England.html

If you use GoDaddy for your web hosting, they offer free photo albums, free email, and free blogs with their domain names. You can purchase website names for under $20.00 per year at this moment. My personal link is:

http://www.godaddy.com/default.aspx?isc=IAPtno100

I recently had a book publisher contact me for a photo they wanted to use on the cover of a book. They found my photo when doing a simple web search for photos, and the title I had for my photo popped up as suitable with the subject matter of their book. Yet they did not buy my photo because the resolution was too low. So here's an important tip for you. Take pictures with the highest resolution you can afford. To sell pictures, you should be in the above 8 megapixel range and at least 300 dpi. Go with 10 to 14 MP if you can. When I had taken my photo, it was back when my camera was only 4 MP. Digital cameras were relatively new, and very expensive. As good as my photo was, it was not good enough for the cover of a book. Lesson learned. My advice applies to contests too. Nikon wants to be able to enlarge your winning picture to poster size. You may not win any contests if your picture does not have a high enough resolution in megapixels.

I recommend that you post your good photos on the internet only in a smaller size and lower resolution than the original. If someone wants your photo, they will come to you for the high resolution copy, which you can then sell. I do recommend that you allow students doing science projects, authors writing books, and people who really have no huge commercial interest to have your photos for free. This gives you free publicity and is just a nice thing to do.

Become an expert. As you get to know your Rheinberg filters and start enjoying them, you might find that you like to take pictures of a particular type of subject. If you build up a collection of good photos, you can consider actually doing some research on your topic and publishing an article or series of articles on your specialty. Your article will be filled with beautiful Rheinberg technique pictures, of course. A series of your articles could have the potential to be a book. A good place to start is Microscopy-UK, or Micscape Magazine, an online magazine for microscope enthusiasts, with articles by the members.

Here is the main link to the magazine:

http://www.microscopy-uk.org.uk/

Here are a couple of links to specific articles like I've described—Rheinberg filters plus an area of specialty:

http://www.microscopy-uk.org.uk/mag/indexmag.html?http://www

.microscopy-uk.org.uk/mag/artjan05/jmcbry02.html

http://www.microscopy-uk.org.uk/mag/indexmag.html?http://www.microscopy-uk.org.uk/mag/artdec03/wdonion2.html

Have you ever thought of giving talks at local libraries, community centers, or clubs? There are many venues for talks and lectures. Do you have your own children in school? Perhaps you would like to give a talk on some topic having to do with the microscope. When my children were in elementary school, I brought my microscope to their classrooms on two separate occasions. The teacher loved the interactive program I put together for her students, which included hand-out pamphlets. To give a talk, you could bring your microscope and some slides, and you would only need a laptop computer and a projector. Presentation projectors have recently come down in both size and in price, making them portable and affordable.

You can actually make money by giving talks about your hobby if you have the right subject and present it at the appropriate venue. As an example, let's say it's the beginning of summer, and the local county buzz in the media is "mosquito control." It's a perfect time to photograph the various stages of mosquito larvae, become a subject matter expert on breeding. Then put together a program for the local chamber of commerce, community centers, or perhaps a city council meeting. You might even get a couple of interviews in the local newspapers as well. These are just a few of the many ways to make extra cash with your microscope.

You can set up a free page to display and *SELL* your photographs here:

http://fineartamerica.com/

I have successfully sold photography on this site, and it is free to display up to 20 pictures.

My microscope interest is in tardigrades. After posting lots of information and photos on my website, I was contacted by a major on-line media producer. It resulted in a video of me that received over ten million views. The name of the video is First Animal to Survive in Space.

http://tardigrade.us/2012/09/04/mike-shaws-interview-on-tardigrades-and-space/

Later, I was contacted by TV GLOBO in Brazil, and I appeared on the number one rated Sunday evening show in Brazil.

http://tardigrade.us/2015/01/04/space-bear-hunter-on-brasilian-tv-globo/

Are you ready to get started?

CHAPTER 2

Microscope Requirements

Before you can take on the task of making Rheinberg filters, you must determine if your microscope can handle them. So let's look at the way Rheinberg filters work.

I'll show you the basics here; however there is a website you can go to and actually demonstrate the Rheinberg effect—the Molecular Expressions website:

http://micro.magnet.fsu.edu/primer/techniques/rheinberg.html

On the above interactive site, you'll not only be able to play with the Rheinberg filters, but you can find out about all other types of illumination, and the myriad things microscopy. There are tutorials that are very hands on, and you'll benefit greatly while having fun.

Before we discuss how Rheinberg works, here is how a specimen looks without it, when you just use a colored filter. This is a tardigrade in *Brightfield* with a green colored filter. You can see the specimen, but there is not a lot of contrast.

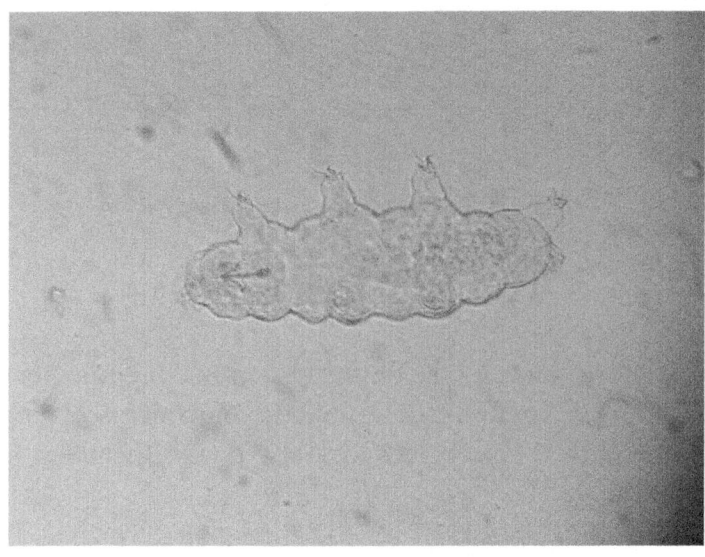

For now, below is an illustration of how Rheinberg works. Look at the diagram to see what the light is doing.

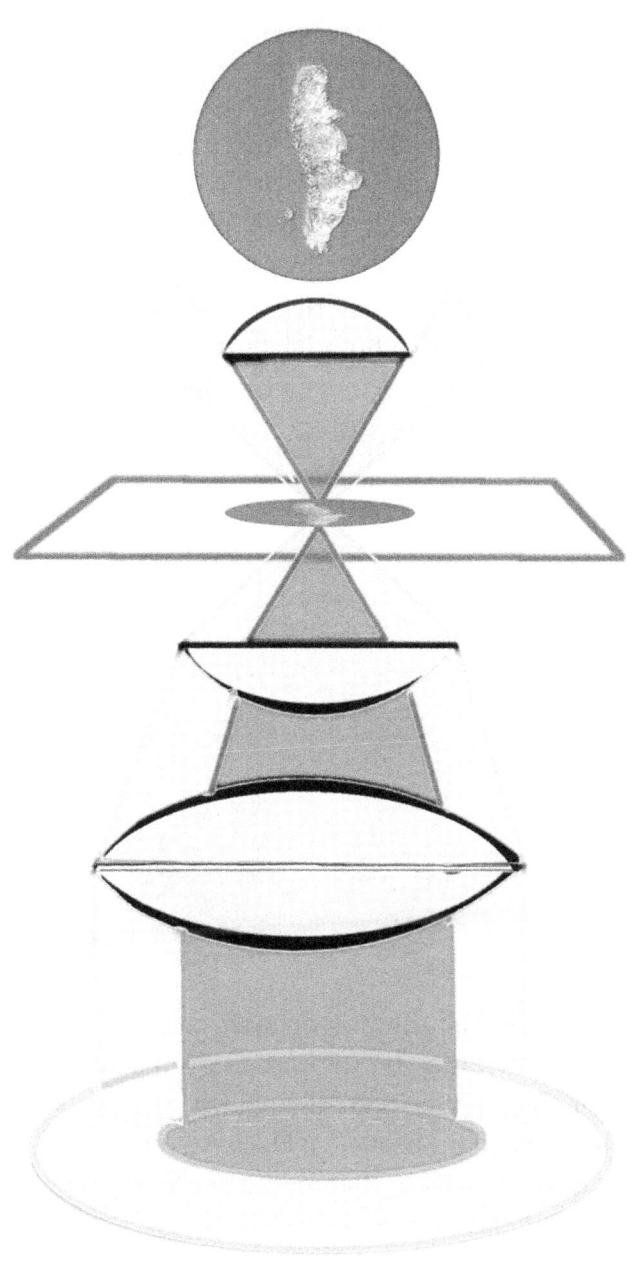

You can see how the colored light produces the result. In the illustration, you can see how the Rheinberg filter is directly underneath the condenser. The light passes up through the condenser- split into an outer ring of pale color, and an inner dark center. At the slide and specimen level, only the outer ring or color illuminates the tardigrade. The dark green center provides the background. At top, we see the view through the eyepiece. You are looking at an actual tardigrade photo taken using Rheinberg technique.

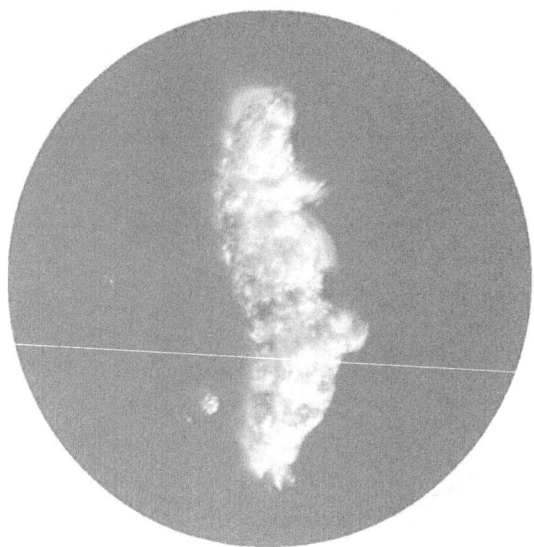

In a nutshell, here is another way to think about how Rheinberg works. If you were to drop a small coin in the center of your daylight blue filter, you would block light. The coin would become a black center stop. A center stop is a disk that sits in the middle of a filter. Its job is to stop or block light.

If a center stop is large and opaque, like a coin, it will block most of the light. Light can creep around the center stop and illuminate the specimen, but the center stop will keep the background dark. This is called *Darkfield* lighting. If a center stop size of 11/16 inch works for your microscope, then guess what? A nice Darkfield filter will only cost you ten cents. Place a dime on your daylight blue filter, and you're done.

When we make a Rheinberg filter we make a special center stop that allows some color to get through, so the background is not black. A coin as a center stop blocks all light. A Rheinberg center stop allows some color to get through. The Rheinberg center stop background is dark, but allows a bit of color to seep through. The Rheinberg center stop is translucent. Yet there is also some light passing *around* this darkly colored translucent center stop. That combination creates the "Rheinberg" effect. Rheinberg is the same as Darkfield, except it allows some color to get through the center stop.

Darkfield and Rheinberg Comparison

Darkfield is where the background, or field of view, is black, and the specimen is brightly illuminated. See a typical Darkfield view below.

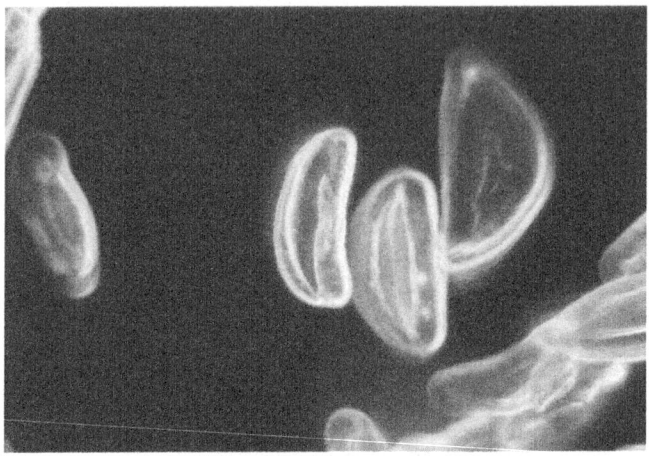

The above picture was made very simply by having a black spot in the center of a daylight blue filter. If I had not used a Darkfield center stop, you would hardly be able to see the pollen grains in the picture. Pollen is practically clear, and very hard to see unless you use some method to give you contrast. Compare it to the picture below using no center stop. It is the exact same slide.

This technique below, using bright light transmitted up through the slide, is called *Brightfield* or "transmitted light."

Your Microscope Hobby

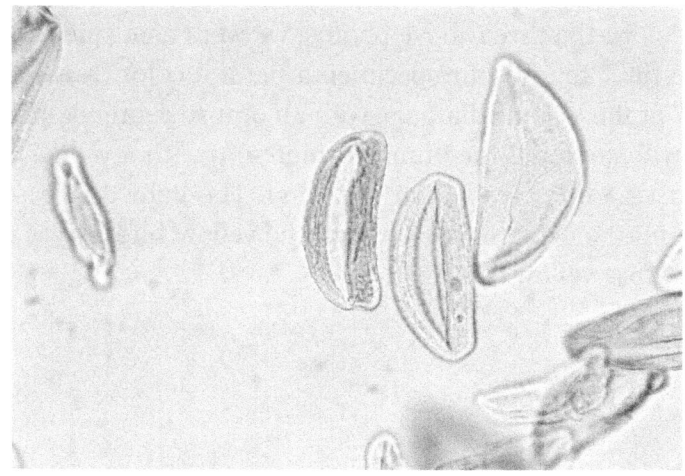

This brings us back to Rheinberg. Instead of an opaque or black center stop, we use a colored center stop, one that is translucent and allows some light to pass through it. We can then have a background, or field, of any color we like. Below is a photo of diatoms, this time, using a red center stop. Remember, although the red color is dark, it allows some light to get through.

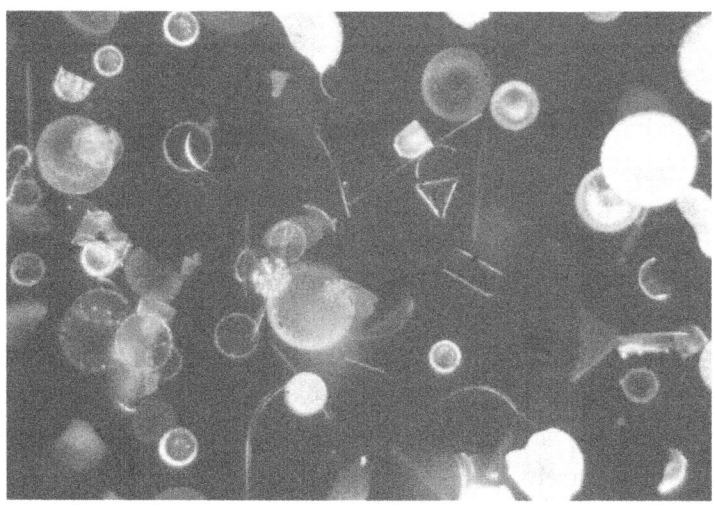

In the three above photos, we were using a daylight blue filter to give our specimen a natural color. Now, let's look at these same diatoms, with an orange center stop *and* we will surround the orange center stop with a yellow colored ring filter (called an *annulus*). The light that passes around the center stop, through the yellow filter, gives the diatoms a yellow color.

That's it. You can use any combination of colors to create fabulous contrasts. This is the Rheinberg technique.

Below are some more examples of specimens, using different combinations. As long as the center stop is nearly opaque, it will give you a dark background and allow some of the color to come through. The subject color is created by the colored ring (the annulus). This provides all of the light for your subject. You can have an annulus ring in any color as long as this outer ring is very light in density, almost transparent.

Your Microscope Hobby

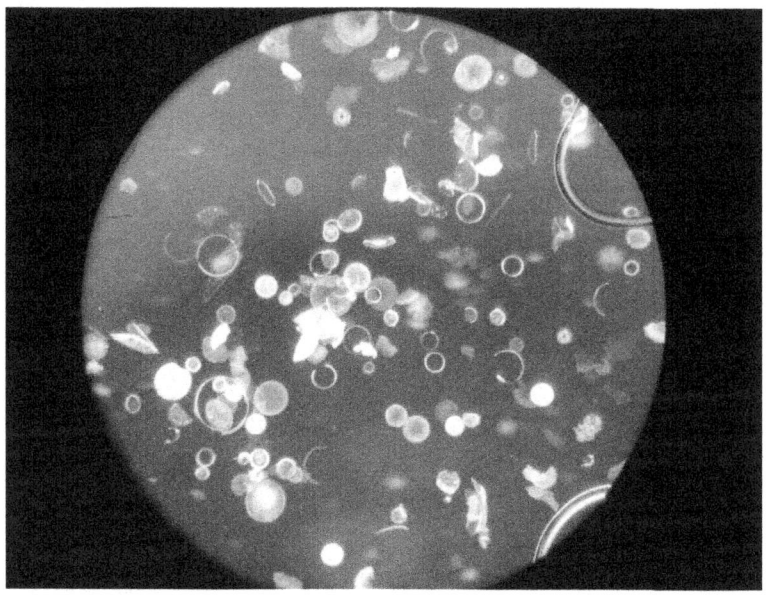

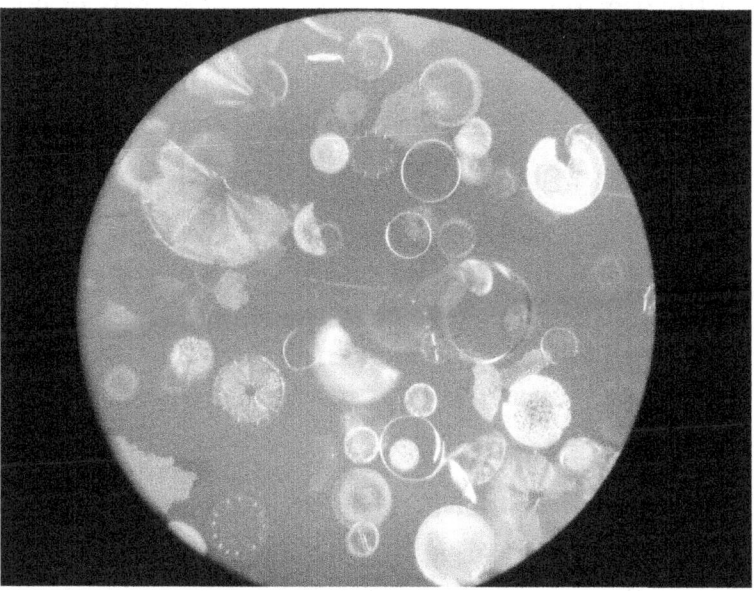

Go back and look at the picture of a tardigrade a couple of pages back, in Brightfield with only a colored filter, but without Rheinberg center stop. Do you see the difference between that and the ones below of pollen? Below left is with a blue Rheinberg center stop. Below center is normal Brightfield (no filters at all). Below right is using red Rheinberg center stop.

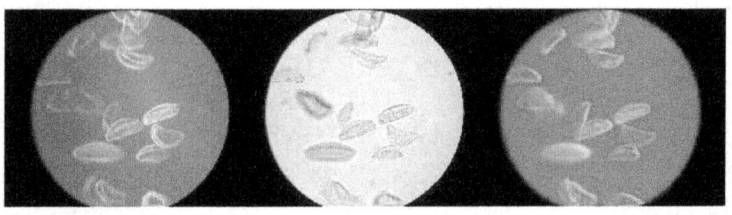

Your Microscope's Requirements

Now let's look at requirements for your particular microscope. Your microscope is really several sets of different optics: eyepiece, objective lens, condenser, and light source. These components will all play a role in how you make Rheinberg filters work with those variable parts of the microscope.

First, Rheinberg works best with an ordinary Abbe condenser or achromatic condenser, rather than a very expensive aplanatic "plan" type of condenser. An expensive "plan" condenser may not work at all. And, you will probably get better results with a 1.9 NA (numerical aperture) rather than a 1.4 NA condenser.

Ditto the objectives. If you are using Neofluor, planapo, or phase contrast objectives, your results may not be as good as if you use ordinary achromats. You are also limited in the power or NA (numerical aperture) of your objectives. Rheinberg filters work best with objectives of 20x and lower. In fact they work best with a 10x objective, and great with a 4x objective. Higher powered objectives (such as 40x and up) require highly precise measurements or a special Darkfield condenser as shown below.

Why is this? Think about the size of the aperture in a high powered objective. It's really small, and we're dealing with a cone of light which we are manipulating with an opaque disk or patch (the center stop). That center stop would have to be extremely precise in size and position to allow only a microns thin ring of light through the tiny opening of a 100x objective. It is like trying to hit a bull's eye with a recurve bow from 100 yards. Even for an Olympic archer this is difficult. Using center stop filters is like using a traditional recurve bow on a windy day. A Darkfield condenser, however, is like using a high powered rifle with a scope—it's an instrument of accuracy, designed for precision. When using higher powered objectives, the center stops we can make by hand usually miss the target.

Let's compare the opening, or aperture, on a 100x to that on a 4x objective. The front lens opening difference is substantial. If your center stop for a 4x objective is not quite centered, or if it is a bit too small or too big, it is good enough. You will get a Darkfield or Rheinberg effect. You cannot have even the slightest error, however, using a 40x objective or greater.

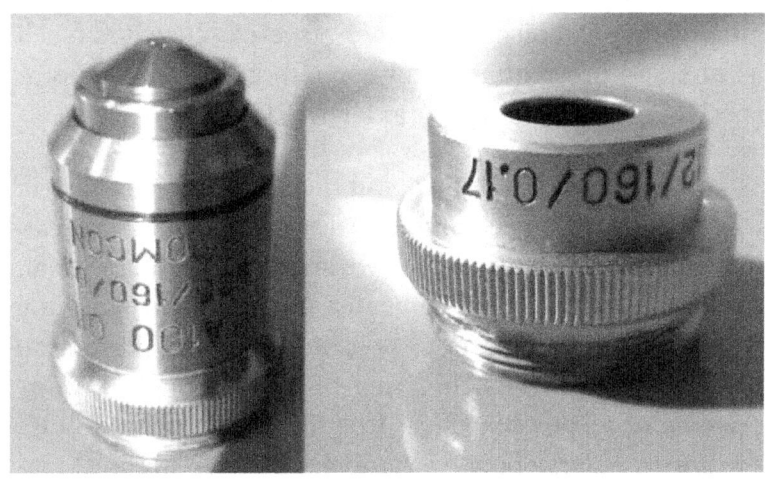

Below is a Darkfield condenser. This does not produce a colored background, only a black background. You could use colored filters to give your subject a specific color, but a Darkfield condenser only provides a black field.

The Darkfield condenser is expensive to buy and messy to use. It won't allow any light through unless you oil the bottom of the slide to it, and of course you must use an oil objective on top. This oil objective must have a built in iris, or a funnel stop placed inside. You are limited in many respects. If you want to see blood cells, then fine, you must go through all that trouble.

The only condenser in existence that uses Rheinberg filters with high powered objectives is the Mikropolykromar. Here is the description from the website mentioned earlier:

> http://micro.magnet.fsu.edu/primer/techniques/rheinberg.html

"During the late 1930s, Carl Zeiss manufactured a special condenser, the Mikropolykromar, designed to produce beautiful Rheinberg images. This condenser is long out of production and is now virtually unobtainable, but it consisted of an aplanatic condenser under which there were three separately controlled diaphragms. The outermost diaphragm controlled the diameter of the field, and two smaller diaphragms controlled the light passing through the central disk. Annular transparent rings of various colors were part of the set. Accompanying these was a set of transparent glass central disks which fitted neatly into the central opening of the annular ring. This ingenious condenser has been used by one of the authors (M. Abramowitz) to produce striking photomicrographs at various magnifications, some of which have appeared in the publications Omni, Time-Life, Scientific American, and National Wildlife."

Now, to continue on the topic of microscope requirements...

We've established how Rheinberg works, the size and type of objectives required, and we've touched on the best type of condenser needed—an ordinary Abbe, NA 1.9, with a swing out filter holder (like the one below). There is still a bit more to know about the condenser.

As a reminder, in order for Rheinberg to work, you cannot have a condenser lens beneath the filter, because that would refocus the light coming through the Rheinberg filter. Therefore, if your condenser has a swing-in lens (like the one above), you must swing the lens out, when use the Rheinberg filter in the holder.

The Swing-In Lens

Why have a swing-in lens at all? This swing-in lens is typically used for lower powered objectives to broaden the light across the entire field of view. This depends upon the illuminator and how its light is focused. In some cases you need to spread the light out when using a lower powered objective. For example, if you are using a 2.5x objective because you want to see a mosquito larva which is quite big in microscopic terms, then you would need to swing that lower lens into place. But when using Rheinberg, you cannot do so. Perhaps a diffusing filter below your Rheinberg filter will help (the white disk shown in the picture above).

In fact there are LED (light emitting diode) illuminated microscopes that do not have the lower condenser lens. This is because the LED provides enough light with any objective. In this case, the filters can be placed either in the condenser's swing-in ring, or right on top of the base illuminator. The preferred location for Rheinberg filters, however, is at the bottom of the condenser. Rheinberg filters can be made for a base illuminator, but there are a couple of possible problems.

One of the problems in placing a filter directly on top of the base illuminator is heat. The Rheinberg filters we make are plastic and can warp or melt if exposed to prolonged heat. If the light source is behind the microscope, however, as in the case of Zeiss and many other scopes, then heat will not be a problem. In Zeiss and other scopes the light goes from the back of the scope to a mirror or prism, and then travels up into the condenser. That light is cool, and will not affect a plastic filter on top of the base illuminator. If the illuminator uses LED light, and the LED's are placed far enough away from the filter, then there is no adverse heat effect. If the heat is not properly controlled, even a glass filter would get too hot and crack.

Some microscopes do not have a swing out ring under the condenser, and instead they use a slider. This allows you to drop a filter into a slider, and insert the slider into the bottom of the condenser. See below picture of a slider that contains a Darkfield stop filter. By looking at the below green filter and the slider, can you guess what the specimen will look like?

Your Microscope Hobby

If you guessed that the specimen will have a black background and a green subject, you are correct. By having a Darkfield stop in pure black, you can simply drop any solid colored filter over it and have a nice Rheinberg effect (without a colored background). This is very useful if you have an assortment of colored filters. Below is how a specimen of goat hair looks using the above arrangement.

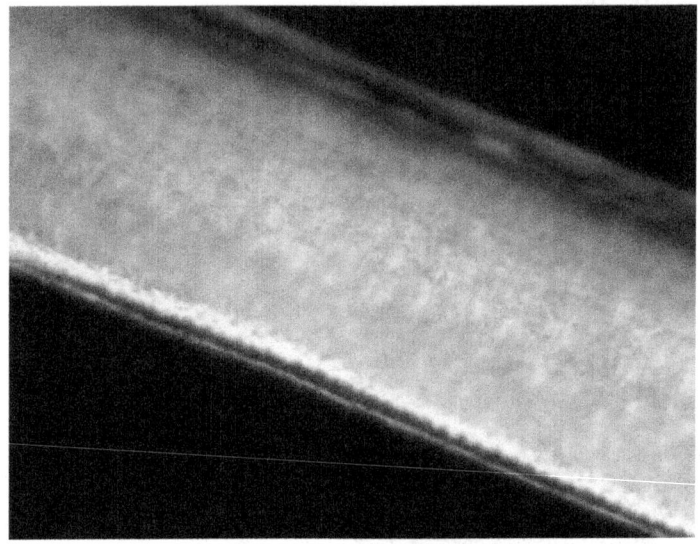

If it were not for the Rheinberg effect, it would simply be a boring piece of hair. Yet, magnification at 25 times shows the texture unique to a goat, giving it a green color against a stark black background, and makes this photo slightly more interesting.

The other important procedure is to keep the condenser iris diaphragm wide open. Just as in the case of the swing in lens, you do not want to reduce the beam of light coming up through the Rheinberg filter. This condenser diaphragm is typically located just above the filter holder.

If you try out a new Rheinberg filter and it doesn't work, double check the condenser iris. You must leave it wide open.

Some microscopes have an iris built into the base illuminator. The aforementioned applies. This base illuminator iris, sometimes called a field diaphragm, must be left wide open. You don't want a narrow beam of light hitting only the center stop of your filter. The light beam must be wide enough to pass around the center stop, lighting up your specimen.

In summary, the key points are: Rheinberg (and Darkfield using a black center stop) works best with lower powered objectives. The condenser and objectives should be ordinary achromats if possible. Position the Rheinberg filter as close to the underside of the condenser as possible. Any iris should be open. There should be no lens between the light source and the Rheinberg filter.

When making filters for others, you will be challenged with all kinds of microscope configurations. Digital photos of the microscope base will be of great help when you can see what you are faced with.

CHAPTER 3

Tools Required

Making filters involves cutting out or punching out a series of disks and rings. These are made of thin plastic, and are fairly easy to cut. In the chapter following this one, you will see a very wide range of materials. The tools you will be using should be able to cut through all those materials.

Measuring

Before we can start cutting and punching out circles, we need some measurement tools. A caliper makes life very easy. A very good electronic caliper can be found on eBay, new, for around $20.00 to $50.00. That's what I use because I make filters for others commercially, and this requires high precision. But you don't need a fancy caliper if you are making filters for yourself. You can buy a plastic caliper in the dollar store (shown here at the bottom of the photo). Look in the tool section of a discount store. You may find one for...a dollar. It's also available in my Amazon store. If you are only making a set of filters for yourself, go the dollar route, since you can experiment and find which filter works best. You will not need the kind of precision you would if you were selling filters commercially. In fact, for a personal set of filters, you can use a high quality steel rule to do your measurements. Pictured are all three options.

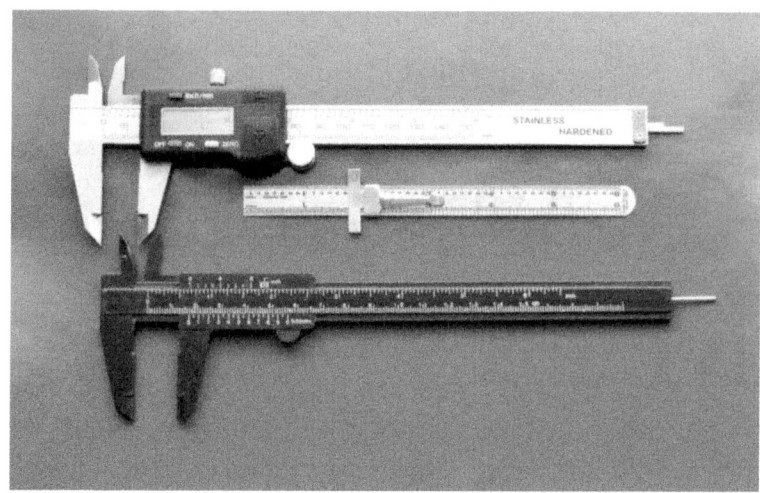

For measuring filter sizes, you can also use a circle template which you can purchase in an art supply store or office supply store.

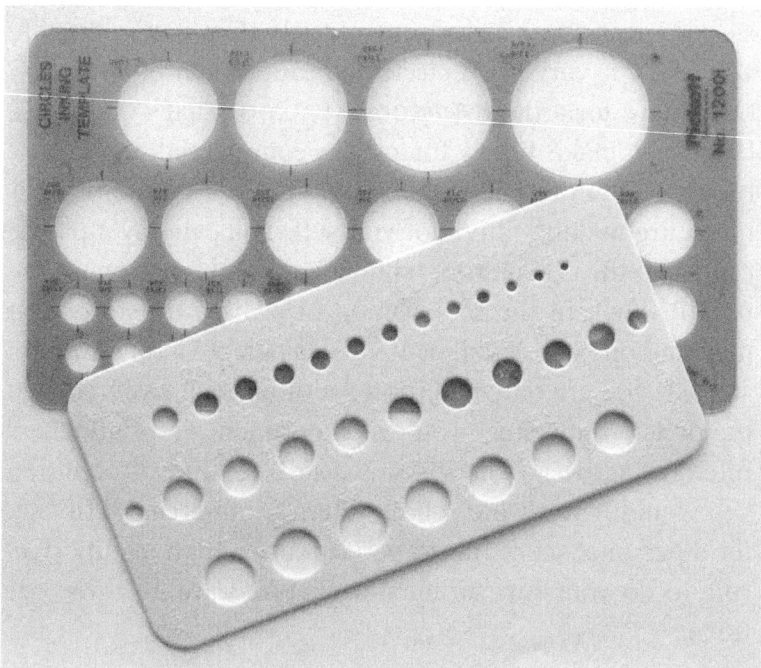

An important part of measuring your work is being able to convert inches into fractions, then into decimals, and back. Sure you can use a computer program, find a calculation on the internet, or use a calculator, but I have found that the best, fastest, and most reliable way, is by using a chart or table. On the back of my steel rule is one such chart, and there are many charts out there on the internet. Search for "metric conversion table," then print it up.

Cutting and Punching

Now that you have measuring devices to tell you what you're doing, you can collect your cutting tools. These tools will allow you to make holes in various materials. It's what's inside that counts, though. Both the hole you make and little circle you punch out are valuable filter components. In essence, we are concerned with making holes, preferably round ones.

Big holes are used for making one thing: Big disks, from which all the filters are made.

Small holes are used for making two things:

- holes in the middle of the big disks (to convert disks into rings)
- center stops (the part that comes out of the hole)

Let's talk about the small holes first. A quick web search for "microscope filter making tools" sometimes results in the suggestion of a cork borer. The idea is that you would use a cork borer to make center stops.

Basically, a cork borer is a tube with one end sharpened, and a handle at the other end. You sort of push and twist through your material to make a perfectly round hole. Laboratories use these to make rubber corks for glassware. These cork borers can be pricey, from $5.00 to $50.00 each.

Wikipedia has a nice explanation and picture of cork borers for the insatiably curious you:

http://en.wikipedia.org/wiki/Cork_borer

And of course you can use Google Images, and see all of the lovely cork borers out there in the geek universe.

You don't need a cork borer to make Rheinberg filters. You can make your own boring tool by sharpening a piece of aluminum tubing with a Dremel® tool if you have the patience. I did so, using old poles from a clothes rack. Here is a rather strong piece of tubing. You can see where I've begun to cut it by scoring around it with a Dremel® tool. Once the length is correct, I'll sharpen the inside and outside of one end, so I can use it as a hole punch.

If you are lucky, however, you may not have to do any work at all. That apple corer in your kitchen may be the exact size of the center stop you need. My apple corer, by KitchenAid™, is 7/8 inches, and using my conversion chart, that's 22mm. I have made a number of filter sets using 22mm center stops, so this would work. An inexpensive kitchen appliance may do the trick.

There is no harm in running to the everything-bath-and-kitchen store, or to your local supermarket to peruse the kitchen utensils for an apple corer. Don't forget to bring your caliper with you. You can look at prices on a couple of these apple corers in my Amazon store.

If you want to make holes in something more substantial than fruit, you may want to try a steel hole punch. These are sold individually or in sets. I recommend you look in flea markets, and smaller hardware stores. They will cut a nice hole in a piece of paneling or plywood, so they are perfect for our thin sheets of plastic material.

By the way, we are still talking about the smaller holes, the ones we use to create center stops and to make holes in the middle of the filter. Most of the hardware chain stores do not carry these steel punches, and if they do, they are very expensive. Smaller hardware stores, Mom and Pop type places might have them, and I've even seen them in dollar stores. They are cheap to make and import from overseas countries, and you can sometimes get a whole set (a *hole* set, ha, ha, ha) for ten or twenty bucks.

Don't bother looking on eBay, though, because these things weigh a ton, and sellers may charge you a lot for shipping. You can view a couple of nice sets in my Amazon store, and at least Amazon has a more reasonable shipping charge.

The sizes I show in the above picture are ½ inch and 9/16 inch. And guess what? I'd say about 80% of the filters I've made fall into this range for the center stop size. Those two sizes are usually perfect for creating the Rheinberg effect with both a 4x and a 10x objective.

Another type of punch is the ready-made plastic craft punch available in the crafts supply store. These are designed for punching holes in paper, so you practically need a hammer to use them on a sheet of plastic. The material can only slide into the punch so far, and you are limited in that way as well, because you must cut your material into strips before you can punch it. You can use this type of hand punch though, and the quality of the punched out circle is decent. I've tried most of the plastic craft punches, and ended up getting rid of them. There are one or two that do work, however, and they are cheap. Look for a size 1 ¼ inch circle – which will give you a 32mm filter size, good for most microscopes.

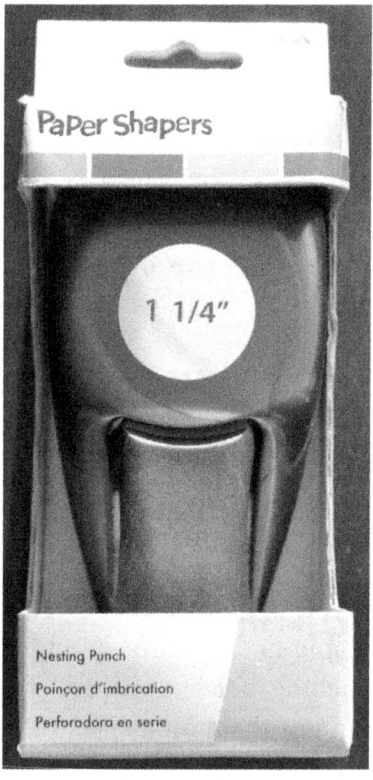

Of all the hand punches, the best circle punch that I've discovered is called the Nesting Punch, manufactured by EK Success®, in their *Paper Shapers* line (above). It comes in a perfect 1 ¼ inch size. I've tested it, and it cuts through paper and thin plastic perfectly. It is excellent for making the outer diameter cut of your filters. It will not cut through a credit card, but it will cut through any of the thinner materials that you use for making filters. Above is a picture of it. You can buy it on Amazon. The punch is operated by using a heavy duty handle, which works fine.

Smaller craft punches will require a mallet.

You will need a mallet to use the small hand punch directly above for cutting anything thicker than paper. Most small craft punches are not recommended for our purposes here.

In order to use metal hand punches and small craft punches (except the EK Success punch), you are going to need a mallet of some sort. Craft stores sell a lightweight version with plastic tips, and the hardware store has a heavier version with rubber and plastic. You can use a wooden mallet as well. For the big steel punches, you must never use a steel hammer. Use wood or rawhide. Here are two of the mallets I typically use.

When using mallets on metal punches, *always* wear safety glasses. Safety glasses are so inexpensive that there is no excuse not to use them.

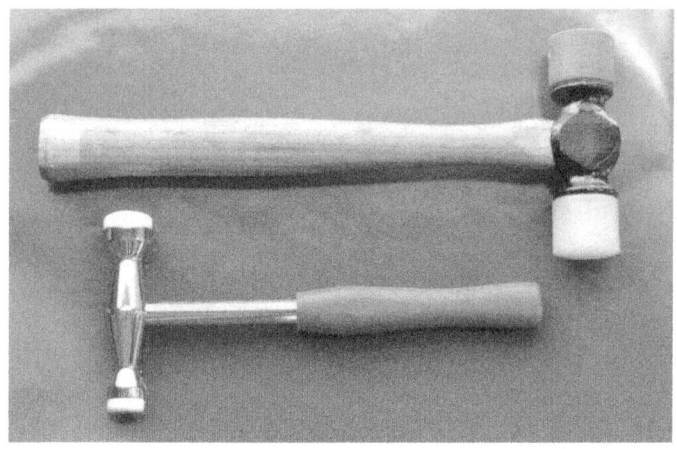

The best method I've found for punching holes is by using a press and a series of precision steel circle dies. The press I use was made by a company called Sizzix®, and these are no longer being manufactured. It is a discontinued item probably because shipping was too expensive. It weighs twelve pounds. It is a wonderful tool, meant for scrapbooking and for making your own die cut greeting cards. You can use this press for cutting out the big filters, and for punching the smaller holes that make center stops and annulus rings.

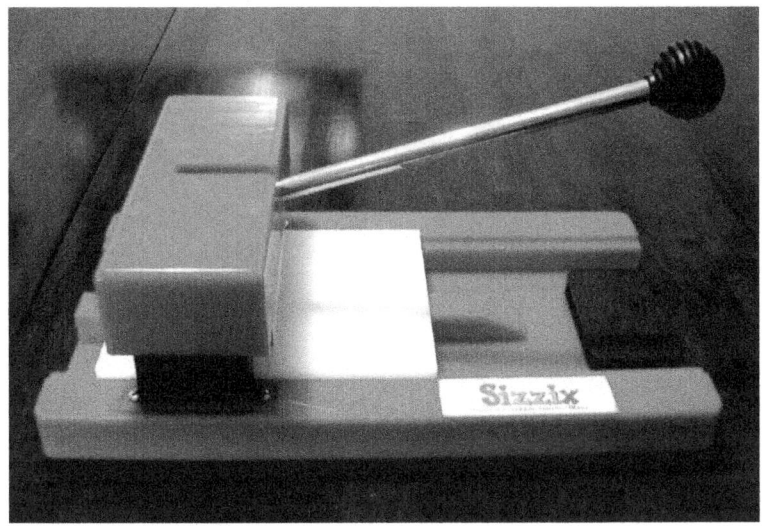

Shown here, it is very heavy and solid. At the time of this writing, they are selling for $15.00 to $50.00 on eBay, and I think the shipping will be in the $20.00 range. When I first saw it in the crafts store, I immediately bought it. I do not recommend the Sizzix® circular dies, because they are not round. You will need to buy an extra cutting pad, because this is a soft plastic that absorbs the precision edge of your die without dulling it. These cutting pads may still be available in crafts stores or on line at the Sizzix® website. You can make your own cutting pads as well, by using cardboard or a plastic kitchen cutting board, trimmed to the right size.

There are a few Sizzix® presses on Amazon occasionally, new and used, but a tad pricey. I added a Sizzix® press to my Amazon store so you can read about it. You cannot purchase it through me directly, but in my store you will see a link to other sellers from whom you can buy one. The Sizzix® press method is the best way to go for making filters. But you may have to do some work to get the press at a decent price, because they are discontinued.

To go with the press, you will also have to buy some low profile circle punches (shown above). You will need to keep your punches at ¾ inch tall or less to fit into the Sizzix® press. These dies are extremely difficult to obtain. These are called *feed-through punches*, or *tube punches*. Typically, they are made by steel rule die companies. If you can obtain these circle dies locally, I encourage you to do so, however, you will find them challenging to find. On the internet, look up companies in your area that make feed through punches or steel rule dies.

Another alternative is to purchase a set of hand punches used for leather working or gasket making. These come in sets. The C.S.Osborne set below has a centering point, which gives you a great deal of precision when making annulus rings. As expensive as the Osborne set is, it's less expensive than buying the Sizzix® press and the equivalent number of punches in the sizes you would get in the Osborne set below. It is a fine set of precision punches.

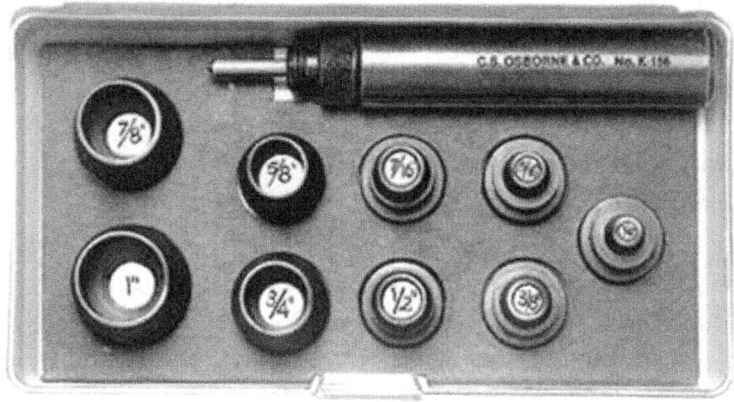

The only drawback is that the centering point will make a dimple in anything you punch. That's okay if you are knocking out the center of an annulus ring. If, however, you are making colored center stops, you will have to place a little plastic protector over the material before you strike the punch.

One of my readers in UK brought to my attention an extremely nice punch set. It is available in UK, and it looks like it would be perfect for making center stops in all sizes, as well as outer diameter annulus rings. It's from the Axminster Tool Centre. Punch sizes in the 29 piece set are: In mm - 2-3-4-5-6-7-8-9-10-12-14-16-18-20-22-24-26-28-30-32-34-36-38-40-42-44-46-48-50mm Compared with the USA Osborne set, the Axminster set is a comparable value. Axminster ships internationally, and you can order it on the following website:

>http://www.axminster.co.uk/boehm-boehm-hollow-punch-sets-prod22429/

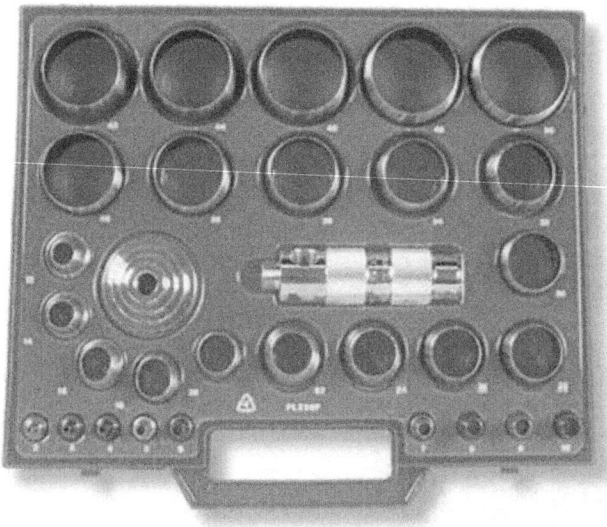

You can also try to obtain the above set from other places in your country locally. A company called Mayhew Tools, for example, sells precision punch sets.

http://www.mayhew.com/

Now that we've covered punches, I want to tell you about another method of cutting circular filters. The compass (not to be confused with a magnetic compass), is an ancient invention dating back to the time of Euclid and earlier. It was extensively used by Leonardo DaVinci, and is a critical navigation tool for measuring on charts, used in present times.

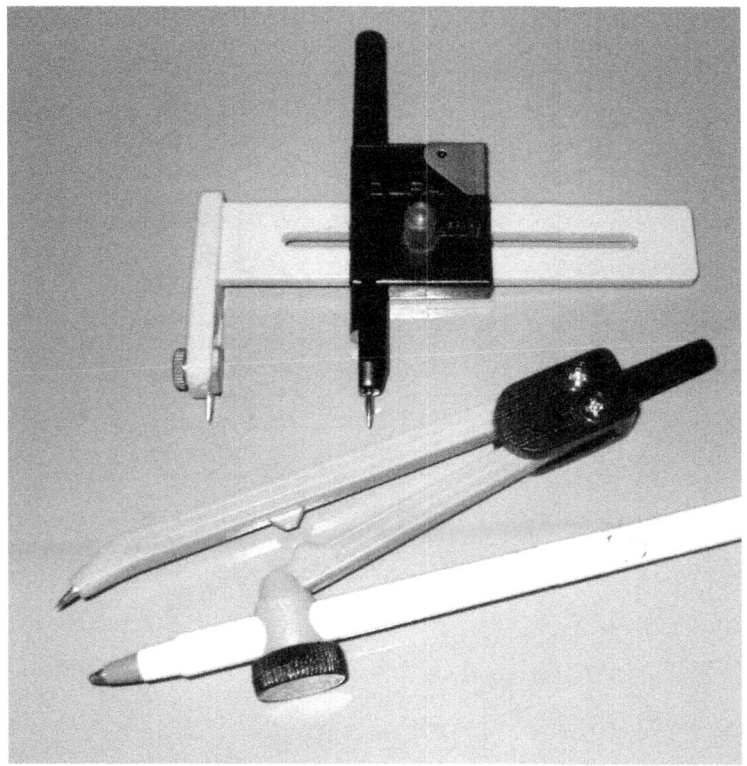

Pictured here are drawing and cutting compasses. You may want to purchase a cutting compass, which is a compass with a built in blade, also known as a beam compass. You can see a nice description of these compasses in my Amazon store. Any good art supply store carries these cutting compasses.

The blades are refillable, and compasses come with spare blades in the handle. Note- the yellow OLFA compass (above) has very simple inch measurement markings on it, unnumbered, so you'll have to make your own precision markings with a sharp scoring tool. The red cutting compass (below) has metric markings, and works just as well as the yellow one.

Your Microscope Hobby

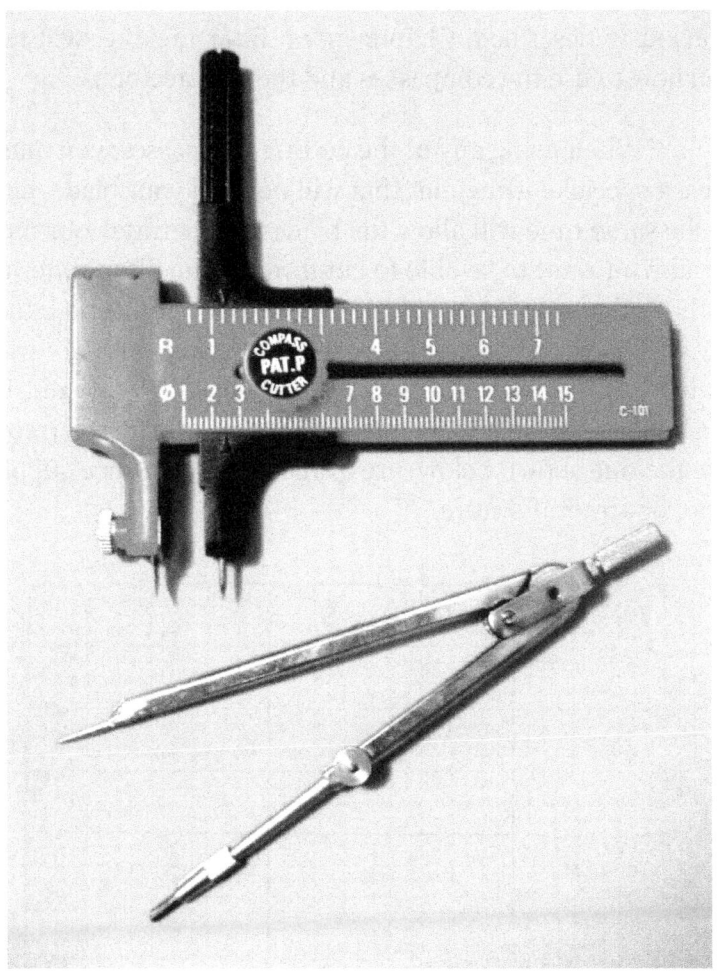

When you purchase a cutting compass, there should be included in the package a little padded disk. This allows you to place the centering needle on the disk, rather than on the surface of what you are cutting. The disk allows you to prevent the compass needle from making a puncture in the center of the material you are working with. These protectors are also useful with the Osborne hand punches previously described. Chapter 5, on filter making, will tell you how to use the compasses and their protectors.

When using any of the cutting compasses, you must have a special cutting mat that will not dull your blade, and at the same time will allow the blade to penetrate your material. You want to be able to cut through the filter material with no damage to your desk or workbench, and no damage to the blade either. To save money, you can get away with using a salad cutting board, the plastic kind sold in the housewares section of a store. Cutting mats (preferred), like the one shown below, are available in any office supply store or any crafts store.

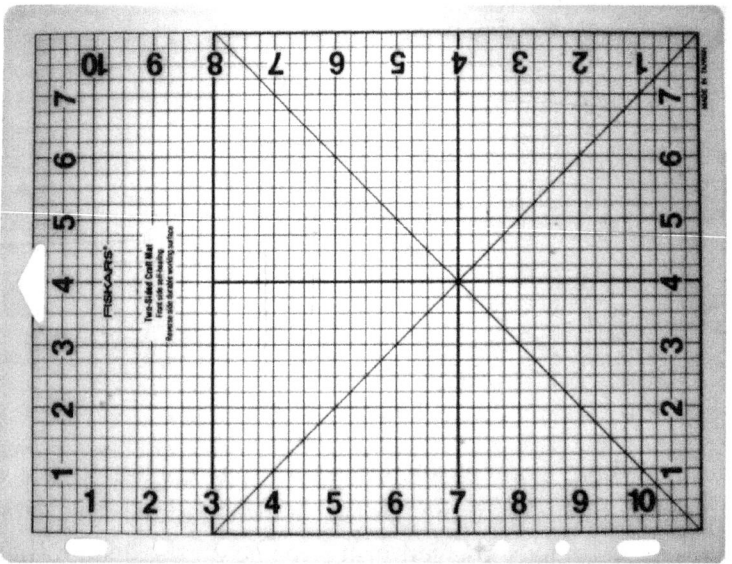

Scoring tools (like the one shown below) are also useful, and important for cutting some of the harder acrylic materials, thicker plastic sheets, or even old CD cases. These are found in the tile cutting section of the hardware store. They are also useful for marking your measurement markings into the surface of steel dies and on surfaces of plastic cutting compasses.

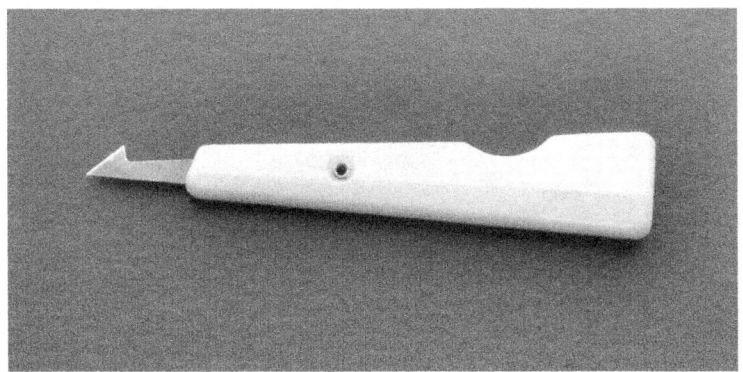

A sharp pair of regular scissors is a must. Get a brand new pair, and a quality brand, and don't use them for anything else but filter making, so they stay sharp. Sharp scissors will provide a clean precise cut, and you need this because under the microscope anything less will make your cut edge look like mountainous terrain.

You will also need a new pair of sharp cuticle scissors.

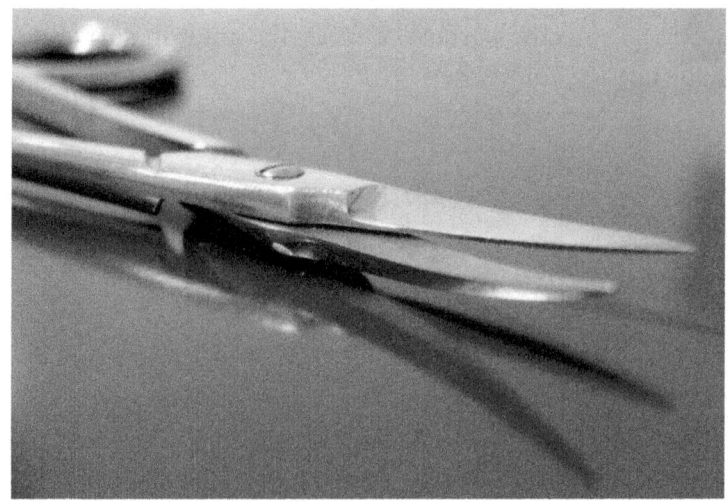

These cuticle scissors have very fine point, which curves along the cutting edge. Cuticle scissors are handy when trimming filters that aren't punched cleanly, and they are good for cutting out test center stops. Get these in the cosmetics or manicure department of your favorite chain drug store.

Finally, there is yet another method for cutting out filters and center stops. A new machine was recently developed to cater to the craft and scrapbooking market. It is called the Silhouette Portrait ®. One of my readers advises that he uses it successfully to make microscope filters, so here is the link:

> http://www.silhouetteamerica.com/shop/si lhouette-portrait

Essentially, you can program the machine to cut precision circles in sheets of vinyl material, acetate, thin colored plastic, even in electrical tape, and in our clear and translucent filter making sheets (Lee and Roscolux). The Silhouette® company even sells its own line of adhesive vinyl in all the basic colors, although you may be better off pricewise sourcing your own. Either way, you now know you can make center stops this way, as well as annulus rings in any configuration. The best news is that the machine is very reasonably priced because it is geared to the large consumer retail market. There are different models of the machine, and it is the model called *Portrait* that is programmable and allows you to make different sized circles. Here is the company's description:

> *"The Silhouette Portrait® is an electronic cutting machine for personal use. Like a home printer, it plugs into your PC or Mac® with a simple USB cable. However, instead of printing it uses a small blade to cut paper, cardstock, vinyl, fabric and more up to 8" wide and 10 feet long. The machine also has the ability to register and cut printed materials."*

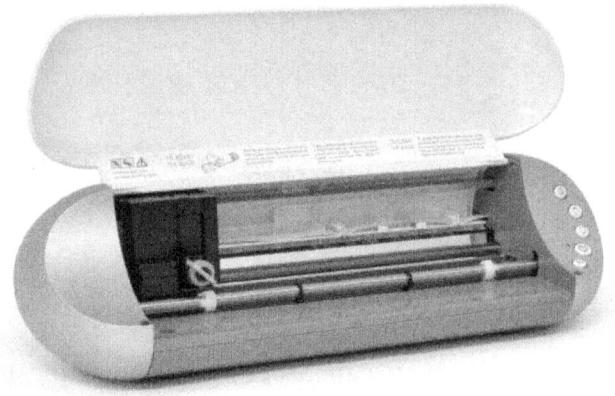

Although I have never used the machine, it has been highly recommended to me by one of my readers. I do intend to eventually buy one and try it out myself. I would appreciate any additional feedback from anyone who uses what looks to be a great new tool for making filters.

Cleaning Materials

When handling materials, be sure your hands are always very clean, because even the slightest bit of dust or lint will show up under the microscope. To clean tool surfaces and even the filters themselves, I use Kimwipes®. I also use them for packing the final filter set into the shipping case. Kimwipes®, made by Kimberly Clark, are available from my Amazon store or any lab supply store on line. You may get a good deal if you buy in bulk from Grainger, and you can call to order them at this phone number, asking for item No. 2017. Phone 800-240-6373. They have a great website, and I've purchased lots of supplies from them, including safety glasses.

Tweezers

While shopping in the manicure department for cuticle scissors, you will need to find some good tweezers. Do not use the tweezers that have a grooved or textured edge. Those are great for pulling out nose hairs, bringing tears to your eyes. But they will mar the plastic of a filter. The cosmetic tweezers that have little flat platforms at the tips are not good either. These are too thick, and when you try to place a thin center stop on a clear disk you will never be able to center it. The tweezers that are bowed at the tips are good for eyebrows, not filters. A pair of thick or bowed cosmetic tweezers will, in fact, get in your way.

The best kind of tweezers? Long and super thin. Ideally, your tweezers should come to a sharp point, as shown in this picture. This way, you can position the smallest thing, and hold objects by the edge if you wish. More about this is in the filter making chapter. You don't need gripping power, you need positioning ability. You are not removing a splinter; you are placing a thin piece of material into a scientifically precise position. Above is my recommendation.

While you can plunk down $25.00 for a pair of professional tweezers, you may want to consider that currently American Science and Surplus has a set of six different tweezers for around $6.00, their item number "88366 TWEEZERS," and within that set you will find what you need. Their stock is limited, however they always carry basic items like this. For convenience, here is the website address:

http://www.sciplus.com

To summarize, each of the tools just described has its use for a different level of complexity desired in your work. My advice is to wait until you finish reading this book, before acquiring the tools I've explained. You will first need to decide which types of filters you want to make, and what techniques you think you'll feel most comfortable with. You may find that you just want to make a half a dozen simple filters for yourself. In that case, you could just use a plain pair of scissors to cut your materials, and you'd only need a drafting compass with a pen to draw some circles. At the other extreme, you may want to go into business, making filters commercially. In that case, you'll want the best tools to satisfy the demanding requirements that customers deserve.

CHAPTER 4

Filter Making Materials

Rheinberg and Darkfield filters can be made from just about anything. You could theoretically rest a petri dish of red wine on top of your base illuminator, throw a thin slice of carrot in the middle of it, and you'd have a nice Rheinberg effect showing a deep orange background with your specimen illuminated in pale pink. People make filters using coins, aluminum foil, and colored sticky dot labels. I have read articles about inkjet printer circles made on the computer to produce very nice Rheinberg filters. I've used many materials, and continue to do so, such as electrical tape, theatrical gels, sign-making vinyl, Con-Tact® paper, construction paper, Easter basket cellophane, CD cases, and old Visa® cards.

You could make a filter with a pepperoni.

In fact, one of the quickest ways to do a test of the size for your Rheinberg filter is to lay a small coin on top of your standard filter, then check for a clean Darkfield effect. We'll get into this, and more, in the Filter Making Process chapter. Many amateurs are quite proud of the metal Darkfield stops they have painstakingly constructed. These consist of a metal disk suspended by three soldered wires in the center of a metal ring. It is something to brag about when the size of inner disk and outer ring is just right, and gives a perfect Darkfield effect.

In this chapter, we discuss the materials that are least expensive and easiest to work with. Although some consider me an authority on Rheinberg filters, I'm not the final authority. My approach quickly shifted from research and exploration to practicality and production. The typical set of filters I sell has about 50 filters in the full set, so I've tended to stay with the same tools, materials, and techniques for the last 10 years. If you find some new materials, a new tool to use, or a new technique, then go with it. Feel free to share it by sending me an email or posting it on my website.

This chapter is divided into four parts.

First we discuss clear materials. You will use these for the platform on which you mount the center stop. We'll call this clear material the *substrate*. You will be making clear disks which are the size of your standard filter. You can use these clear disks in a variety of ways, but mainly for mounting the center stop and for supporting the annulus ring.

Second, we talk about the material for the center stop itself. That is the little colored patch that goes in the center of the Rheinberg filter. That center is pasted or embedded right in the middle of the clear disk mentioned above.

Next, we'll talk about materials for the annulus. The annulus is the colored ring that surrounds the edge of the filter. It rests on top of the clear disk. This annulus is made so the inside of the ring is flush up against the center stop.

Finally, we discuss the sticky stuff you use to hold the components together. This could be glue, the adhesive backing of the component, or lamination.

In this picture, there is a slight space between center stop and annulus for illustration only, just so you can see the clear disk. In reality, there is no space. The center stop and annulus are flush against each other. The annulus surrounds the colored patch in the middle very tightly. The picture here is just showing you the three components.

The Clear Disk

Clear materials are abundant, and what you will need is something that is optically clear yet flexible and thin. A trip to the office supply store yields a bonanza of report covers, presentation sheet covers, and so on. My favorite material is called a presentation or binding cover. These come in packs of 25 or 100 sheets, and are bit pricey, but the quality is very good. Make sure you don't accidentally buy the frosted kind. You want something firm, which you can cut or punch holes in, and as clean and clear as possible.

You can also buy individually a few standard report covers which have clear covers, and this may cost you less if you are only planning to make a few filters. A one page-sized sheet of clear material will give you about 25 clear disks, if you don't waste too much. Therefore, if you're just making a small set of filters for yourself, then you may want to start this way. You can probably buy a three pack for a bargain price, and you're done with this component. You can skip to the next section of this chapter.

If you want to go for super high quality, you can buy what is called *Duralar Polyester Drafting Film*, from any good art supply store. It will set you back at least $5.00 for one 18 x 24 inch sheet. One website describes it as "unsurpassed by other films for clarity, strength, stability and flatness. Dura-Lar Polyester Film will not tear or absorb moisture, assures perfect registration of layouts and color separations, features archival quality, consistent color and heat resistance."

Finally, you will want to consider laminating. A great clear material is lamination plastic. It is also possible to laminate the center stop right into the middle of your filter. You don't need to buy a laminating machine because you can have your office supply store run a sheet of hot lamination material through their machine. How could you do it? You could lay several center stops of construction paper in a raw sheet of laminating material. Laminate, and punch out disks.

In the above picture, you can see examples of center stops laminated, and some of the circle holes where they were already punched out. Suffice it to say that the lamination material itself acts as a transparent window for the center stop. Then you would simply lay the annulus ring on top. More about laminating in the filter making chapter.

The Center Stop

Construction paper, or colored paper is a very good material for Rheinberg filter center stops. There is a fantastic book called *Exploring with the Microscope*, by Werner Nachtigall. The author suggests using colored paper for Rheinberg filters. I actually started out making filters this way. It's a great book that covers every aspect of what you do with the microscope in a clear and simple way, even though it was written before the age of digital photography.

Here is a picture of some of the colored papers you might find on line and in stores. Some are glossy, some are matte, some thick, and some thin. Your best bet for filter materials is to shop at discount and craft stores. The crafts store is a great place to get construction paper and other colored papers. Look in the scrapbooking section too, as many scrapbook pages are sold individually, and they may have just the color you want. Don't forget gift wrap, but buy it after the holidays at 50% off.

You would attach the colored paper to your clear disk using cement or glue, or by laminating—all explained a little further down.

Our next center stop material eliminates the need for cement, glue, or tape. Avery® makes sticky labels in the shape of colored dots. Some of these are translucent plastic, and all you have to do is layer them. Some are made of colored paper.

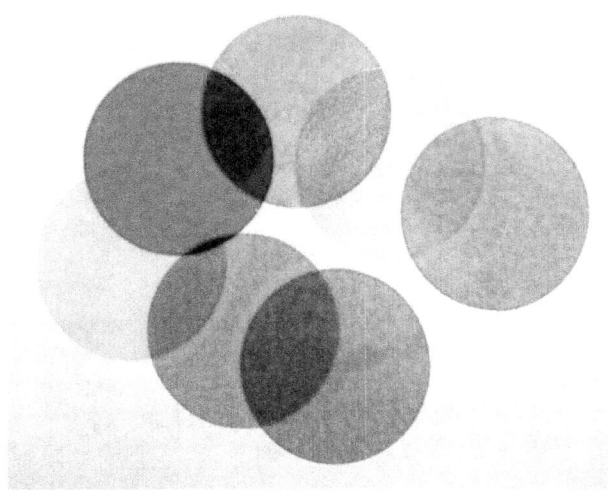

In essence, the three primary colors allow you to have all other colors. In fact, green as a separate color is included so you can work with that as well. You are able, therefore, to create orange and violet, as well as the few tertiary colors of blue-green, red-violet, and so on. These dots are a nice solution, but the one big drawback is that they are not permanent, so eventually they will peel off. You could, however, spray coat your plastic filters with clear acrylic, and make them more permanent, or you could laminate them. Although you cannot use these commercially, because of the peeling factor, for personal use they may serve your purpose well.

My favorite material for making center stops is electrical tape. Besides black, this comes in many colors, but you will have to look for it. Look in the paint and automotive sections of the hardware store, and look in smaller hardware stores. It is available in ½ inch, ¾ inch, and even in a 1 ½ inch width.

In some dollar stores, you will find a five-pack of colored tape in a brand called SA, and another brand now put out by Duck Tape. 3M™ Scotch ™ has a series as well. Do not buy the kind of tape that has fibers in it. Make it a habit to look in hardware stores and smaller discount places for electrical tape. I've built of quite a collection of colors this way. The colors often look almost the same, however they are slightly different in hue, and this makes a noticeable difference under the microscope, so buy all brands you can find. Here is a part of my collection, and I have the various shades of red and blue, as well as odd colors like violet and orange.

The advantage of these electrical tapes is that they stick nicely to the clear plastic filter material substrate, and as center stops they are fairly permanent. These center stops can withstand heat of about 80C (or 176F), so your illuminator light should be no problem. This material is what I use in all of my commercially sold Rheinberg sets.

Colored electrical tape also available in my Amazon store as well:

> http://astore.amazon.com/mikesmicroscopestore-20

There is one slight disadvantage to using rolls of tape. You are limited in your center stop size by the width of the roll. Thus, if you must make a center stop that is wider than ¾ inches in diameter; you are limited in variety of colors. Although 3M™ Scotch ™ rolls do come in a 1 ½ inch width, you can only get red, yellow, green, blue, and black. So ideally, you would like to find more colors in large self-adhesive sheets.

A new product that is out by Scotch 3M™ is called *Expressions* tape. It is a thin tape and comes in all imaginable colors. You can double layer this tape and use as you do the electrical tape.

Colored self-adhesive sheets are available in some art supply stores. Look in the vinyl stick-on letters section. The manufacturer C-Thru ™ Better Letter, makes colored stick-on letters and they also make solid colored sheets. These do work well, and I have successfully used them for center stop material. These materials occasionally come in several colors, but you will usually find only black or white. Buy the black sheets for Darkfield and Oblique filter making. It is perfect for this use. A good artist's supply store will be your best source for these sheets in colors. Utrecht Art Supply is a good online source. You'll have to call them to see what colors they have, because on the internet it looks like they only carry white, which they call "plain." Here is their website:

http://www.utrechtart.com/index.cfm

A good alternative to the individual sheets is Con-Tact ® vinyl. Some hardware stores and several websites on line sell rolls of colored Con-Tact® in any color, but you will be forced to buy a whole roll. Con-Tact ® is made by Kittrich Corp.

I have a few basic colors in my Amazon store, but you'll find more colors available from a dealer called Hardwarestore.com, in Westminster, MA. These rolls are heavy, so shipping is slow and pricey. Link:

http://search.hardwarestore.com/Search.aspx?query=contact%20paper

If you do buy one of these rolls, I recommend you start with the black Con-tact® paper (shown here), because you will make a lot of sample test filters, and an occasional set of Oblique filters. You need black for that. My roll, which is 18 inches by 75 feet, will last you about ten years. I double layer the black to make absolutely sure it is opaque.

A less expensive way to go is with colored pieces of adhesive backed plastic which you will find in a sign-making shop. You will find a bonanza in one of these shops. Many of the nicely lettered signs you see are made from letters punched out of this vinyl. What you need to do is go to one of the nearby places that make all kinds of signs. Ask if they have any small scraps of material that they are throwing away. In my case, they wheeled over a trash can full, and I picked out the tail ends of a dozen colors.

You will quickly find out that not all of these colors will allow enough light to pass though, so you will have to do quite a bit of testing until you hit upon the right color that makes a good center stop. Don't forget to peel off the vinyl from the protective paper backing when you do your test. Stick a small piece on clear plastic. Then hold it up to the light. Old CD cases work as a good test surface. What I do is actually make up sample filters, and then see if they work. Obviously, the best way to test your material is to put it in your microscope filter tray, and turn up the power in your illuminator. You can also can use a flashlight, or an LED light that you can carry with you. Test each color to see if light passes through it.

In testing material for a center stop, it has to be dense enough to provide a dark background in the color you want, yet translucent enough to allow a bit of light through. This is a trial and error method to find what works best for your microscope. If you are outdoors, or if the store you are in is too brightly lighted, the flashlight exercise is futile. It will be difficult to make an evaluation. Don't forget that you can always double the layers of a material to add density.

The Annulus

This is the outer part of the filter, ring shaped, that gives the subject its color. Here, as a quick reminder, is our filter diagram.

The annulus allows light to pass around the center stop, and illuminate the specimen. Therefore, this material must be practically transparent. It should be colored just enough to provide tint, like the gels they use for the stage in the theater and in Hollywood for lighting the actors and the stage or set. In fact, one of the very best materials you can use for Rheinberg filters is the professional lighting gel. But first, I'll show you some materials that are less expensive and more readily available.

Back to the office supply store, and a quick stroll down the aisle to "report covers" will take you to a bonanza of annulus material. These report covers come in all types of material, and in all price ranges. Here is a typical report cover, actually more of a paper holder.

You want to buy something that is thick, but clear with a nice color tint to it. In other words, don't get anything that is frosted, or cloudy looking. Those may look nice on a report, but you do not want the frosted effect dimming your microscope illumination. Look at the pictures here for some ideas. Ideal thickness for all of your filter materials is .007 to .008 inches. You don't need to be that precise, because you'll soon find out that many good materials are .005, and some are .009. The red report folder shown here is .006 inches, and is a bit floppy, but will work fine for filters.

The other report covers shown here are .009 inches, very firm, but tend to be a bit cloudy in color. They will do, but you are better off with a clearer material, even if it is thinner.

A very inexpensive material, called Decco Film, is what the florist shops use, and what is typically used to make gift baskets. This material is an extremely thin colored plastic that comes in rolls of 100 feet by 24 inches wide for under $10.00 per roll at this writing. One roll yields a huge amount of material, so plan on making all your own gift baskets from now on.

You will have to laminate this material between clear plastic for it to be really useful, but it is a very cheap source of transparent color. You can buy this material from ULINE:

>http://www.uline.com

ULINE sells this cellophane or polypropylene in colors of blue, green, red, pink, purple, and yellow. I found the equivalent called "Decco Film," at an art supply store on line. If you check out the art stores in your area such as Michael's, A.C. Moore, or Jo-Ann Stores, you will find many useful filter making materials.

On the subject of cellophane, periodically, you can get some from American Science and Surplus at www.sciplus.com in small packs of 8 ½ x 11 inch sheets. Look in their section called materials. The have some really nice shades in their package, and again, plan on laminating it, because it is so thin.

Another type of colored film or plastic sheet is called Clear Lay® Sheet, and available on line from Grafix Arts. The thickness is .005 inches, so that's a tad thin, but still very nice for filter work.

>http://www.grafixarts.com/product/ClearLay

That brings us to the next material, Hollywood and theater lighting gels. Each gel sells for about $10.00, and you get a huge sheet of about two feet square for that price. There are literally hundreds of shades and colors to choose from.

Although I have purchased a number of these sheets for making larger filters, there is a really great thing you can do for free.

The companies that make the gels have sample filter packs, in a width of 1 ½ inches (38 mm). That sample size is perfect for 90% of the filters you will make. A few microscopes use filters in the plus 45mm range, but Zeiss is 32mm, Lomo is 33mm, and most others match that standard of between 32mm and 34mm, including newer Chinese scopes. So with one free sample gel pack, you have every imaginable colored filter! Some places do sell these packs (for about five dollars), but most of the time, you can get them for free as long as you don't load up on them during one visit. Be fair, and just take one or two packs. It will last you a very long time. Where to get these sample packs?

If you are an educational institution or a student, get a free swatch book on the site.

http://www.leefilters.com/lighting/packs.html

Roscolux has a good website that shows you every one of their colors, and spectral analysis if you want as well:

http://www.rosco.com/filters/roscolux.cfm#colors

Go to Roscolux's "Where to Buy" link and find a supplier near you.

You can just walk into the store and ask for one or two swatch books. You may even buy a full sized gel sheet of a particular color you like. There is a beautiful pea green color that creates a perfect pond life background, and 1 ½ inches of the stuff in a swatch simply may not be enough material for you! Buy a sheet. There is a marvelous peach color that makes a daphnia look like Daphne, and a straw color that makes a hydra look like a scarecrow. Using the filter book, if you are careful, you can punch out two filters per swatch book page. I got my filter packs at a local theatrical store. I was a regular at the store when I lived in the area, and bought all kinds of filter making supplies there.

Here are some more on-line suppliers of these gels:

http://www.stagespot.com/

http://www.cheaplights.com/cart/page31.html

They have a sample pack under $5.00

http://www.premier-lighting.com/sales/colorfltr.html

The easiest way to get started with gels is to get a sampler pack by Victor Smith called: Smith Victor Color Effects Rainbow Filter Pack with Six 12" x 12" Gel Filters.

You can get these in my Amazon store or at Adorama Camera on line. Go to:

http://www.adorama.com

Type in the search box "Victor Smith Color."

Note—these are gels, and this color selection is great, but they are thinner than the Clear Lay material mentioned previously. You will need some support substrate like a clear disk underneath when using these for annulus rings.

Having the widest possible variety of colored plastic sheets will give you many options that go beyond Rheinberg annulus rings. Therefore, it is important to have the maximum choice when it comes to colors. You can use colored sheets of material for both solid disks, and for annulus rings. Solid disks come in handy when you use reflected light, or what is also called incident illumination. Here is an example of how useful this can be used.

If you want to take a picture of salt crystals, in natural light, and you'd like to add a colored background here is all you do.

Throw a solid colored filter into your filter holder, and sprinkle some salt crystals on a slide. At first, when you look through the microscope, everything will be green, including the crystals (above). Here's a trick, though. Illuminate the crystals from the top, using a bright white LED light, or a fiber optic cable, or even a bright hi-intensity desk lamp. You can even use colored LEDS, or a colored filter in front of the incident light (say your desk lamp). Once you have this set up, you can simply switch in other colored filters to change your background (see below).

Below is some clear nylon braided fishing leader, as seen through the microscope, without any special background. Then, look at it with colored LED lighting from the top and a solid teal colored filter.

Your Microscope Hobby

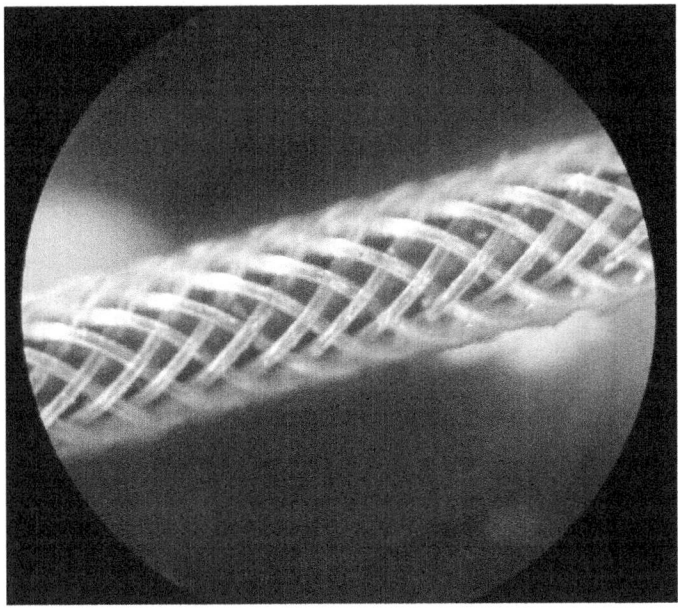

It's not exactly the Rheinberg effect, but you can make some very nice images using only solid colored filters, and colored lights from the top.

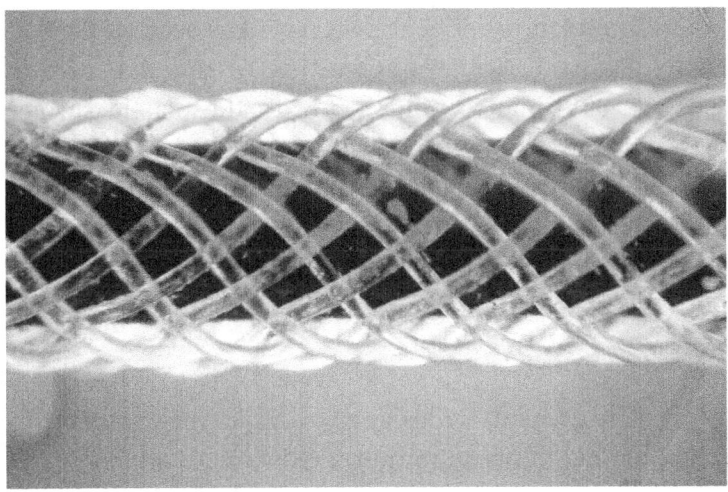

I'm sharing this technique with you because Rheinberg filters do not always work in every microscope, and may not work with certain objectives or condensers. So if you have one of these challenging situations, you can still achieve beautiful effects by simply using a solid colored filter from underneath, and a separate light source above the microscope. That's what the above pictures demonstrate.

You are now familiar with the three components of Rheinberg filters: The Clear Disk (a support substrate), The Center Stop (for the background hue), and The Annulus (gives the specimen its color). With a large variety of materials to choose from in each category, we are almost ready to tackle filter making. First, however, we look at some other materials that will be useful to you in the process.

Cement, Glue, and Lamination

To attach paper center stops to the center of a clear plastic disk, you will need some type of glue that will dry clear and continue to hold firmly to the plastic surface. Some glues are strong and will actually melt the plastic, warping it. Other cements dry quickly and are all-purpose, but they release from the plastic when dry. The best type of glue I have found for this work is *fletching glue*, which is an archery specialty cement for attaching the feathers to arrows.

This cement is designed to hold the feather to a carbon arrow, to an aluminum arrow, or to a wooden arrow. Therefore, this cement bonds various materials very well. Arrows take a lot of abuse. They often go completely through targets, and sometimes you have to pull them out the other side of a hay bale. This really challenges feathers not to come off. The brand I use is "Fletch Tite," from The Bohning Company, Ltd., Lake City, MI 49651. It comes in a little ¾ oz. tube. You can use other brands as well. Most fletching cements come in a tube that has a needle-like nozzle that stays clean, and that makes it perfect for cementing precision filters. You would only need to apply a drop in the center of the filter, and press the center stop into place. You will get the hang of the right amount of glue with some practice. You can buy it from my Amazon store, or from an on-line supplier, 3Rivers Archery at:

http://www.3riversarchery.com/

Call them toll free at 866-732-8783.

You will find that this cement has all kinds of uses. You can actually assemble some very creative segmented filters by laying your colored component filter pieces on top of a clear sheet. Carefully, you draw a bead of glue along each seam to join the pieces. The glue dries clear, and then you punch out the new filter from this composite. More on how to do this in the Filter Making Process chapter.

You can also use double stick tape to hold paper center stops to the clear disk. In the scrapbooking section of the crafts store, you will find this double stick tape.

When laminating, I only use hot lamination because I sell filters commercially, and they have to last a very long time, undergoing a great deal of usage. I simply cut out my materials and bring them to Staples® to run through their hot laminator. There are cold laminators for home use, however, I've never used them.

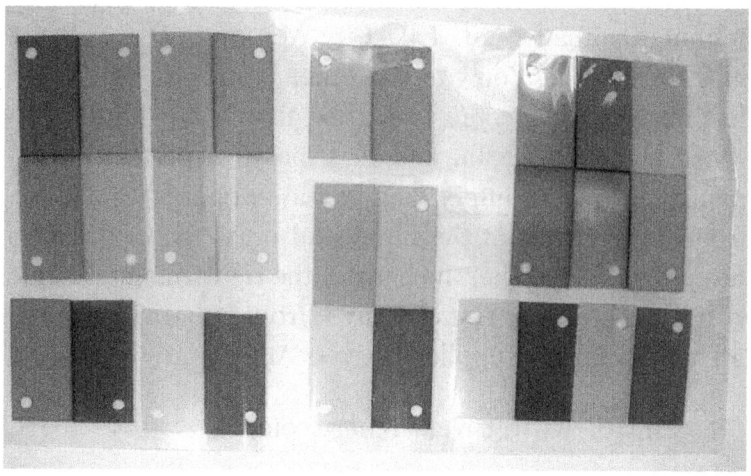

Pictured here, is an example of a hot lamination I've made, prior to cutting out the filters. Where you see two or four colors together, that is the preparation for making multicolored filters. You would simply cut your circles across the dividing lines in the middle of two colors to get multicolored filters.

Other Materials

Craft Foam is an abundant and inexpensive material sold in crafts stores. Get a small sheet of each color, and you can use it to make spacers when you stack your filters. Craft foam (pictured below) provides a lint free protective surface, which you can use to line your various filter cases.

Your Microscope Hobby

Save old credit cards, used gift cards, plastic hotel room keys, and those advertisements that arrive in the mail printed on laminated card stock. These types of plastic cards are good for making test disks when you want to determine the outer diameter of a filter.

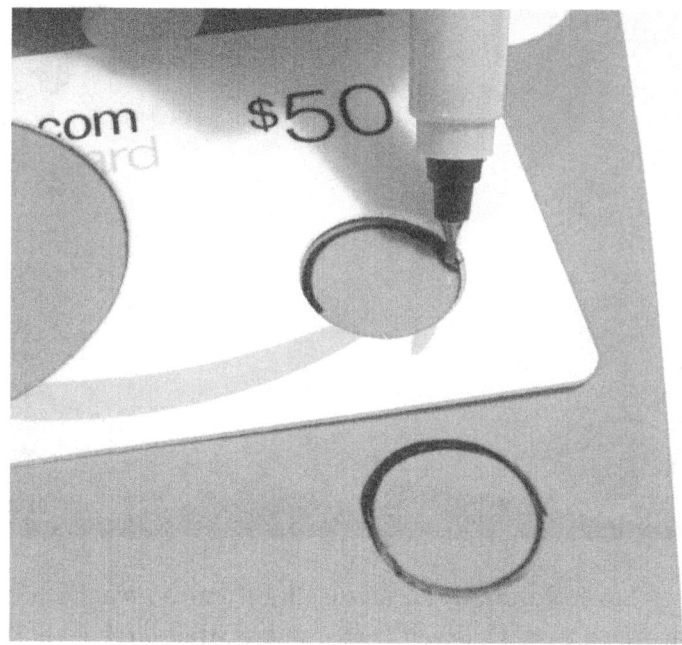

Since these plastic cards are opaque, you can also use them to make filters for oblique illumination (covered later in this book). They also make nice templates for drawing circles on filter making colors.

Do not throw away old CD's or plastic CD cases. When you burn CD's or DVD's and make a mistake, do not throw away the useless disks. CD's provide a good working surface when you need to punch out or cut filter materials, and you don't want to scratch the surface of a table. I use old CD's when using my punch press, if a particular circle die is too tall to use with the standard protection pad.

Below pictures of microscopic crystals were made using old CD case covers. How?

Your Microscope Hobby

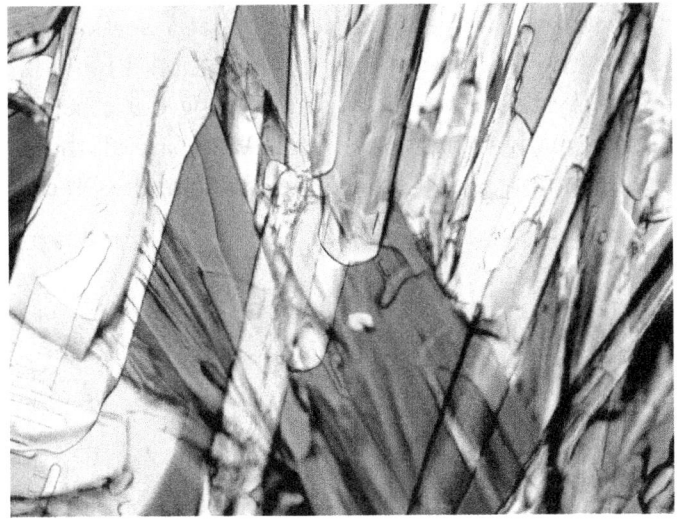

Clear plastic CD cases are very useful as interference filters when using polarized light. All three pictures here are of the same section of the slide.

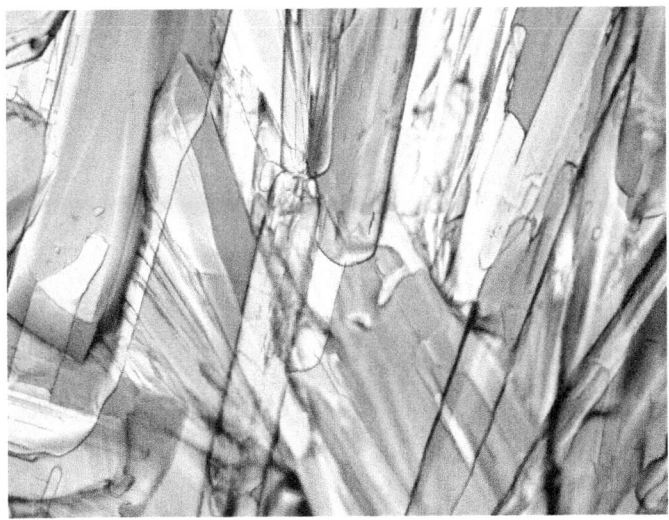

CD cases will give you a myriad of beautiful colors as a background, and also produce the opposite colors in your specimen. This is explained in a later section of this book. For now, compare these three pictures I took of Epsom Salt crystals. Simply by rotating the CD case, I was able to get many variations in color. They are all the same specimen, simply with a different position of the CD case.

CHAPTER 5

The Filter Making Process

Before you can start cutting, punching, and assembling, you must know what the correct sizes of filters will be for your microscope and condenser combination. First you must determine with precision the outer diameter size. Next, each of your objectives may take a different center stop size filter, so you must also figure out the best center stop size for your filters. You can do this easily by making some test disks.

Since this process is based on general filter making skills, I have incorporated the test filter procedure right into the general filter making process.

What will the test set reveal? Not only will the test set determine what size your center stop should be, but this also equals the size hole you need inside your annulus ring. They are the same. What size hole would you make in the middle of your colored disks? The test set will reveal this. For each objective you use, you would usually need a different size center stop and matching annulus ring.

This test set of about half a dozen clear filters with different size black center stops will allow you to see which center stop works best for a particular objective. If you can achieve a nice Darkfield effect, then Rheinberg will work perfectly. This exact size works with a particular objective.

A 4x objective will usually need a different size center stop (and annulus ring hole) than a 10x objective will.

NOTE – Stated earlier, you can use Rheinberg filters with all objective sizes up to about 40x. Using an objective of 40x and above, you need a great deal of precision, and you normally cannot simply use a center stop filter to give you Darkfield or Rheinberg effects.

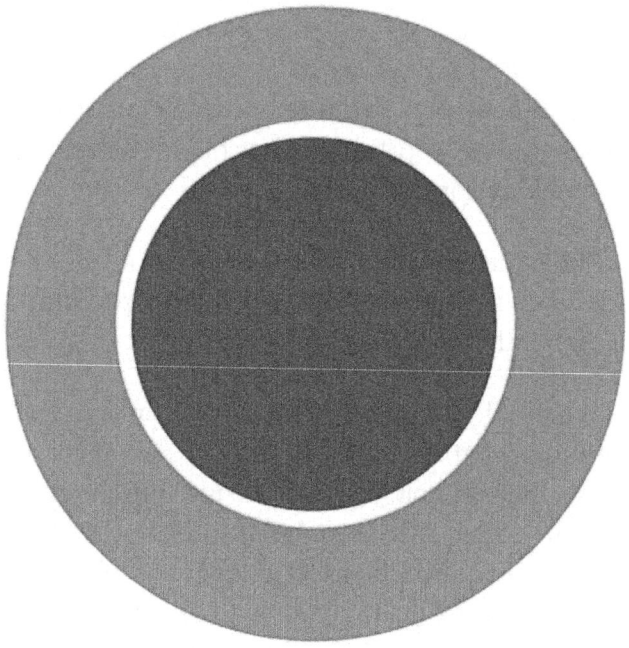

In this illustration, there is a clear border between center stop and annulus- but that's for illustration only. In reality, the center stop is flush with the inner diameter hole in the annulus ring. The measurement is the same. If you have a 5/8 inch diameter center stop, then you cut a 5/8 inch hole in the disk to make an annulus ring.

If you are lucky, you may be able to use a 4x and a 10x objective with the same size center stop. But you won't know this until you make your test set. Can you use a 20x objective and a 4x objective with the same size center stop? Probably not.

To summarize--only by making a test set of filters will you know exactly what size center stop works best for each of your objectives. You will then know what size hole to make in your annulus ring, and what size center stop to make in various colors.

Outer Diameter Sizing

Many of the people who order filters do not always know the exact outer diameter size of their filters. Often, people buy used microscopes which have no manual, or microscopes that do not include filters. The difference between 31.5mm and 32mm may not seem like a great deal, however it is a major problem if you have a scope that takes 31.5mm filters and you've thought all along that it takes Zeiss standard 32mm filters. People tend to think that Lomo and Zeiss are the same standard, however 33mm is the correct size for Lomo scopes.

For this part of the process, I use old credit cards, hotel key cards, rewards club cards. Here is how such a disk looks. In the picture, I'm also using it as a template to trace my clear test disks.

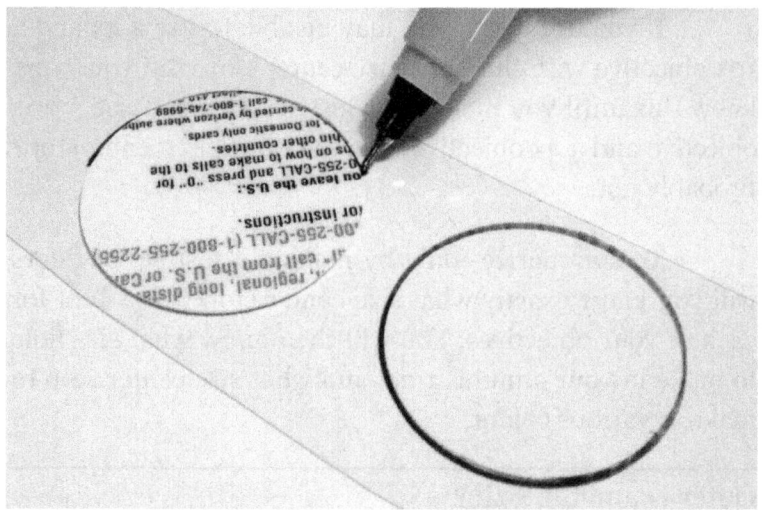

The best material for determining the outer diameter is a solid disk from the firm plastic material. Use a set of large punches, and you are done.

If using a circle cutter, the trick is to go very slowly, making a lot of passes, not pressing down too hard. If you rush, or push down too hard, the blade will bend and distort the size you think you are cutting. When cutting the card, you may find that the compass cutting blade shifts, and you do not get a precision circle.

Use a precision metal ruler or your caliper to set the initial blade distance on the compass cutter. Tighten the thumbscrew tightly. And a brand new blade is a must. Take your time and go slowly until the circle is cut. Work on the protective cutting mat described in the Tools section, or use an old CD as your cutting surface.

Since you are only making a test disk right now, you can also consider an easier way of doing this. First, you can simply score the circle by making a few passes around the diameter with the blade. Then use a sharp scissors to cut out the disk. You will get a fairly clean disk this way, and you can use a nail file to smooth any irregular spots. Of course if you want precision, you can make many slow passes around the circle until the blade cuts completely through. This dulls the blade, though, and you will need to replace it sooner. Remember—at this point, it's just a disk to test the outside diameter of your filter holder.

When using a compass cutter, you must make sure you follow up and measure the circle you've cut using a good caliper. Drop or slide the disk into your filter holder to make sure it fits nicely; make sure it is not too tight and not too loose. If incorrect, cut another disk, adjusting accordingly until you have a perfect fit that is just right. Measure your final result with a caliper, and write it down on the disk itself.

After you have the perfect size disk that works in your scope, then use a thin permanent marker to mark the exact location of the setting on your compass cutter. You will actually make a line on the compass cutter that shows the precise location and setting for your outer diameter.

I recommend that you use the scoring tool to score a deep groove in the plastic along the edge of the desired setting. Then, you can use a black marker with fine point to fill in the score line and save that setting.

Whichever compass cutter you are using, you should follow this procedure. Trial and error determines the size of the outer diameter. A caliper measures the final result. The compass cutter and the disk are marked permanently with that result.

Making Clear Disks

Now make a set of about ten clear disks in the exact same outer diameter size. These can be cut by using punches or a compass cutter. You can use a compass cutter, because it doesn't matter if there are dimples or holes in the center of these clear test disks. You will be covering up those dimples with black center stops. A compass cutter with a couple of passes over clear thin material will work fine.

Since these are test disks, you can even use a scissors to cut them out. Once you know the correct size, you can use a circle template or a drafting compass with a scoring tool to mark your circles. You can simply draw circles on the clear material (above photo). Then cut them out by hand with a scissors. These filters are for testing, and they simply need to be precise enough to fit well in your filter holder.

When making filters commercially, however, that's just not acceptable. You can't charge someone for precision filters and cut them out with a scissors. Even test filters for a client must look professional and be precision cut. More on that in the business section later.

Making Center Stops

You can see how to the set up strips of colored tape for punching or cutting. This is how to mount all of your colored electrical tapes so you can punch them for filters. Keep your strips to about eight inches in length.

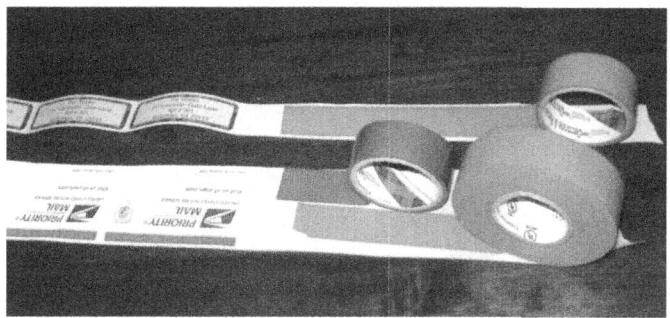

Find a waxed paper sheet or strip from the backing of any adhesive label sheet. A trip to the post office will yield plenty of this, because Priority Mail labels are on a roll, and after the labels are used, the backing is laying around available to you. Many of the postage labels printed by the postal clerks have a non-stick backing strip as well. Just ask the clerks for some of this. You will soon have a stockpile of your own non-stick backing for colored tape strips.

Once you have some non-stick backing, simply unroll some colored tape, and cut it into a few 6 to 8 inch strips, and then apply them to your non-stick backing.

Center Stops for Size Testing

Now you must cut out a series of center stops to test the correct center stop size. Making test filters is a two step process. Step One, you cut out clear disks as explained above. Remember, clear disks are the substrate. Step Two, you make the center stops using black, or any opaque material.

Since you might need a different size center stop for each size objective, I recommend you make a set of about 6 test filters. When I sell these commercially, I make about 8 sizes of test filters for the client to try.

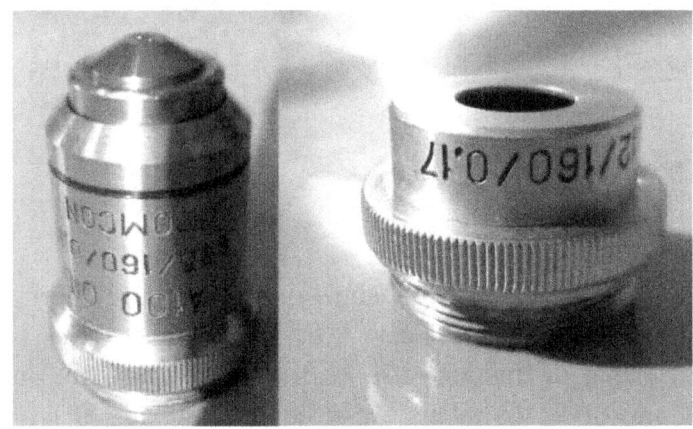

Do you remember from Chapter 2 (pages 29 and 30) why center stop filters do not work for 40x objectives and higher? Although this was covered it makes more sense now if you take another look at the photo above. Think about it by comparing the size of the lens on each objective. You've got that tiny little lens on the end of a high powered objective (pictured on the left side of the photo). Knowing how difficult it is to cut out a plastic disk with precision, do you seriously hope to cut a disk precisely enough to block the light going into that tiny opening, all of it except for a very miniscule perimeter (which we call the annulus)? It's a lot easier to do when you've got a larger lens opening to work with like that of the 4x shown on the right side of the picture.

Your Microscope Hobby

Below picture shows how cutting a center stop with a compass cutter leaves a hole or a dimple in the center stop.

If using a compass cutter, you do not need to worry about the dimples or pin-holes in the middle of your black center stops. Patch the hole to make a completely opaque center stop, by just hand cutting a tiny piece of black material and patching it over the hole.

Using a compass cutter to make center stops is a very difficult challenge. Above is a photo of the Sizzix® press and a precision punch, where I'm punching out a set of test center stops.

There are alternatives to the Sizzix® press and specialty punches. You can use hand punches, which are available in different price ranges. These are described in the tools section. They work absolutely fine for making center stops. Don't forget about the Silhouette Portrait® machine.

The centering Osborne punches, however, will leave a dimple or hole in the center stop (see below), so after you punch, stick a tiny patch of center stop material over that dimple.

Now, go ahead and punch or cut out black center stop circles in gradually increasing sizes. Below is a chart of the sizes that you could start with. You don't need to use these precise sizes, as long as you log the size of each center stop, and mark the center stop filter accordingly.

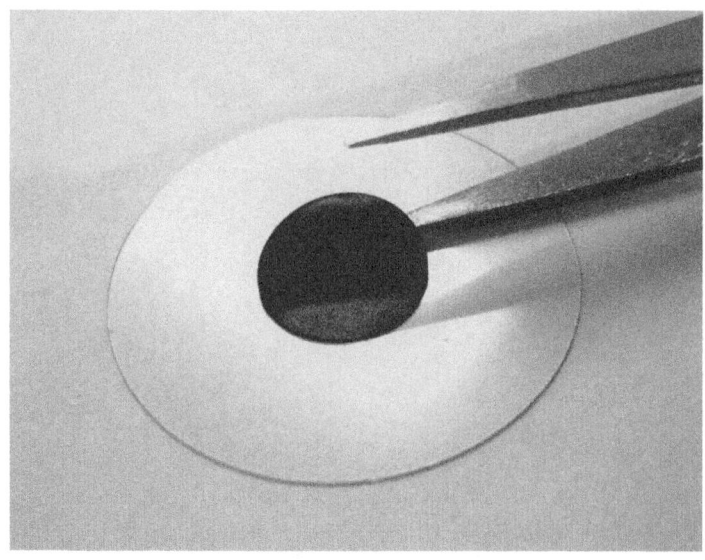

After you cut out the center stops, peel away the backing, and using a tweezers, lay the sticky black disks on the clear filters.

Don't waste time trying all kinds of special techniques to center these. Nothing works as well as placement by eye. I've tried using an expensive steel centering ruler, with numbers starting in the middle and going outwards to both sides. I've tried spinning devices, like a turntable to see if the center stop moves when you spin the disk. Nothing mechanical works. You've got to learn to center these stops by eye. If it looks centered, it is.

Your Microscope Hobby

Flip the filter over, and stick the label on the back-right under the center stop area. This is critical. I use white labels to mark the individual center stop filter sizes. Keep track of each center stop, so you don't mix them up. I recommend that you actually line up the material you have punched out, and one by one, mark each filter accordingly: A, B, C, D, E... etc., using white labels. You can cut small pieces of any white label to do this. I actually punch out my white labels using a 10mm punch.

You will end up with a set of Darkfield filters. After you have your test set assembled, you will need to test each filter with your microscope. You want to systematically test these Darkfield filters, and you can use the test instructions below.

Testing Your Filters

What follows are my testing procedure instructions, and you can provide these same instructions to your customers:

Set up a slide of diatoms, pollen, cotton fibers, or butterfly wing, and focus and get it as perfect as possible in Brightfield—no filter. Start with your 4x objective. The condenser iris must be all the way open.

In general, the condenser should be racked up as far as possible. Racking the condenser up and down is not something you should have to do a lot. Always leave it all the way up, if you can, and then maybe rack it down a tiny bit for a particular specimen.

When you have a nice looking specimen with your 4x objective, focused in Brightfield, try each test filter, one at a time, without changing any other setting. Do not try to focus on the filter plane. Leave everything as is. All you are doing is testing each Darkfield center stop size with the 4x objective, which is already perfectly focused.

If the background appears gray, then you don't have the best filter size for that objective. When you find a center stop filter that gives you a black background with your specimen in bright white across the entire field of view, this is the correct center stop size for your objective.

The background must be black, yet you must see the specimen across the entire field of view in bright white. At this point, move your slide around, or even change specimens to another type of material. You want to make sure that across the entire field, you will have brightness, yet at the same time the background remains black. The only way this can happen is if the size of the center stop is matched exactly for that particular objective.

Once you have determined the correct center stop size, you can mark your test chart. You would make a check mark under the 4x column, in the row where that size test filter is located. Or just write down - "Filter letter X works with 4x objective." Don't rely on memory. Write down your findings.

Next, go through the same process with the 10x objective, and so on. For each objective, you are looking for the best sized center stop that gives the Darkfield effect with that particular objective. You will wind up with a list of your objective sizes, and a corresponding list of which center stops work best with each objective lens.

Again, don't try to make the center stops work by adjusting focus or racking the condenser. Set the specimen up focused in Brightfield, and then test the Darkfield center stop filters by switching them in and out.

When will Darkfield center stop filters fail to work completely? They usually don't work if the filter placement is too low beneath the condenser. Some microscopes have no condenser, or a condenser that is modified so it has no filter ring holder.

Can you place a filter on top of the illuminator? Filter placement on the illuminator would be okay to change the color of the light; however it is not very compatible with using center stops. In some cases, yes, you can use a very large center stop and place this on top of the base illuminator. It might work.

When placing a filter on the illuminator, most of the time the distance between filter and specimen is too great, and the light path cannot be controlled precisely enough. And there is also a potential heat problem. A base illuminator can become very hot, unless the light source is LED, or the light source is behind the microscope and channeled up the base through a series of mirrors. Heat will warp these plastic filters we are making.

The following is a sample testing chart. You will need make your own chart, based upon your hole sizes, or punches. Simply modify the chart I am providing. Whatever you do, you'll need to write down the results of your tests with various objectives and center stop sizes. Once you have a list of your objective sizes, and a corresponding list of which center stops work, you can begin making a final set of filters for one objective at a time.

Stop Size	4 x Objective	10 x	Other
1/2 inch		X	
5/8 inch	X		
3/4 inch			

In the above example, we are making a set of filters to go with your 4x and 10x objective. You have determined that the 5/8 inch center stop works best the 4x. The 1/2 inch size works best with the 10x. You will now finish the last two steps in making your set of Rheinberg filters: Create the annulus rings, and make the center stop filters.

Colored Disks

Now, you can finally start on the fun stuff. You are going to begin Step A of the process of making annulus rings. In this step, you begin by cutting out the colored disks. You already have established the exact outer diameter in the previous step. Get out your colored sheets, and cut yourself a set of disks, in every color you have. It is your choice whether to use a circle template and cut with a scissors, or whether you cut with a compass cutter, or simply score a circle and cut with a scissors. If you use a Sizzix® press, you can do this very quickly. If you are using a compass cutter, these disks will have a slight dimple in the center. It's okay for annulus rings. If you use a set of large precision hand punches, you will not have that dimple problem. The dimple in the center goes away in any case, when we get to the next stage. Here is a set of colors with the dimples showing.

You could, theoretically, use the little center protector to avoid making the dimple in the center. You would have to use clear tape, however, to hold it in place, or it will slide around as you spin the compass. But you don't need to use a center protector when making disks for annulus rings, because you will be removing the center of the disk anyway. The only case where you would need a center protector is when making the clear disks is when you want to make solid colored filters. Depending upon your variety of materials, you should wind up with about a dozen disks in all colors.

Annulus Rings

If you used a compass cutter to prepare the colored disks, then each one already has a dimple or a tiny puncture hole in the exact center. If you used a punch of some kind, then may be no center dimple. Either way, you will need to use a 5/8 inch punch or set your compass cutter to make a 5/8 inch circle. (Remember – we chose 5/8 arbitrarily above. Use the size that works for you.)

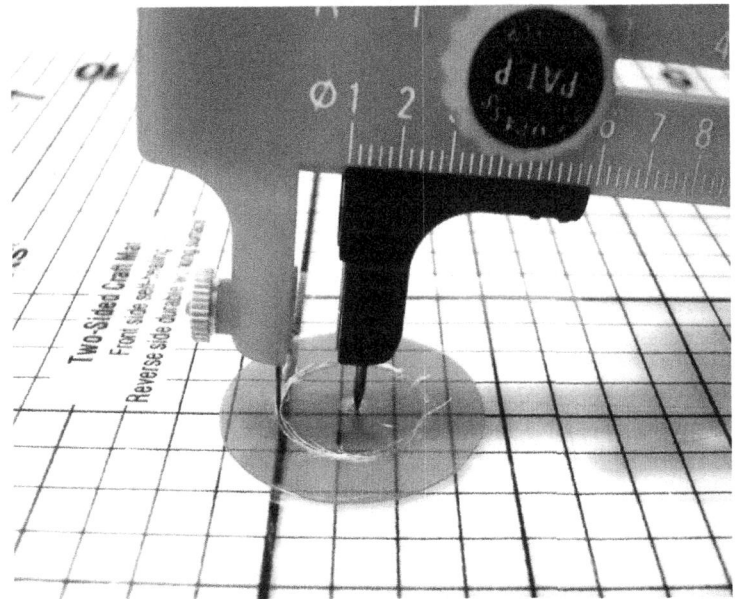

If using a punch, then center this punch in the colored disk, and make the hole. As you will see, that dimple is gone and you are left with an annulus ring.

If you have dimples in the centers of your annulus rings, and you are planning to use the Osborne punches, this is a marriage made in heaven. Those dimples actually give you a really precise way to cut out the hole, because you only need to set the Osborne punch in the center of the dimple. Bang! Your hole will come out perfectly centered. It will be a clean cut. If you are using a hand punch, then you may have to go through several filters before you get one annulus ring with a perfectly centered hole.

You can discard the center piece, no matter how you went about cutting it. If the inside edges of your annulus ring are rough, you can smooth them with a rat tail file or trim any frayed parts with a cuticle scissors. You want to wind up with an annulus ring having an inner diameter exactly the same as your center stop outer diameter.

Repeat this process until you have a complete set of annulus rings in all colors. You see below both the rings and the solid colored filters that were cut out at the same time (except they were not turned into rings). You are now 2/3 done with your Rheinberg filter set.

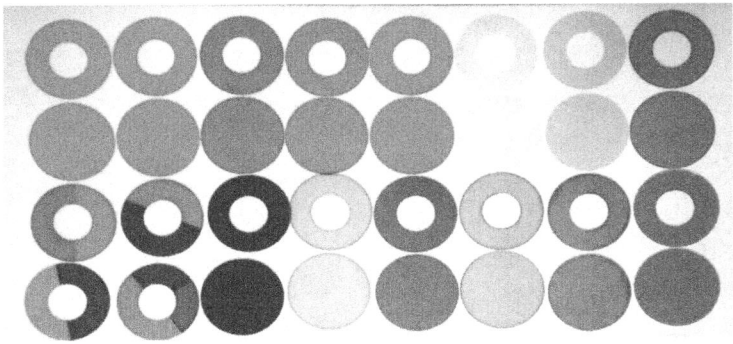

Final Colored Center Stops

You can go ahead and cut out center stops in the same exact size of the annulus holes—in all colors. Remember, your test set determined the size of the center stop. You should have a variety of colors in plastic tape, and hopefully some scrap sheet of signage colors. If you are using rolls of tape, in 90% of cases your center stops will not exceed the ¾" width of the tapes. You should be in good shape to obtain most colors using rolls of plastic tape.

If by chance your centers stops and annulus holes are over ¾ in, you'll have to go with the wider more expensive rolls of plastic tape, and rely upon sheets of signage colors.

Let's start with the hard way to cut out center stops. If you use a compass cutter, this will now be a major problem. The point of the compass will make a tiny puncture in the middle of your center stop. This is unacceptable, so you will have to use the plastic protector which prevents dimples and holes.

When using the protective disk, you'll have to secure it with tape. And you end up cutting through this tape, and it is really tricky to keep the protector in place without it sliding around. That makes the process really cumbersome. If you are making these filters for yourself, however, then fine. It's a one-time pain you'll have to endure. There is no way you'd want to do a set like this commercially, as it would take too much work for the investment of time.

There is, however, a much easier way. You may remember from the Tools chapter that we showed a circle template. That is what you can use to draw quite a precise sized circle on your center stop material. Use permanent marker, and then cut out the center stop with a cuticle scissors.

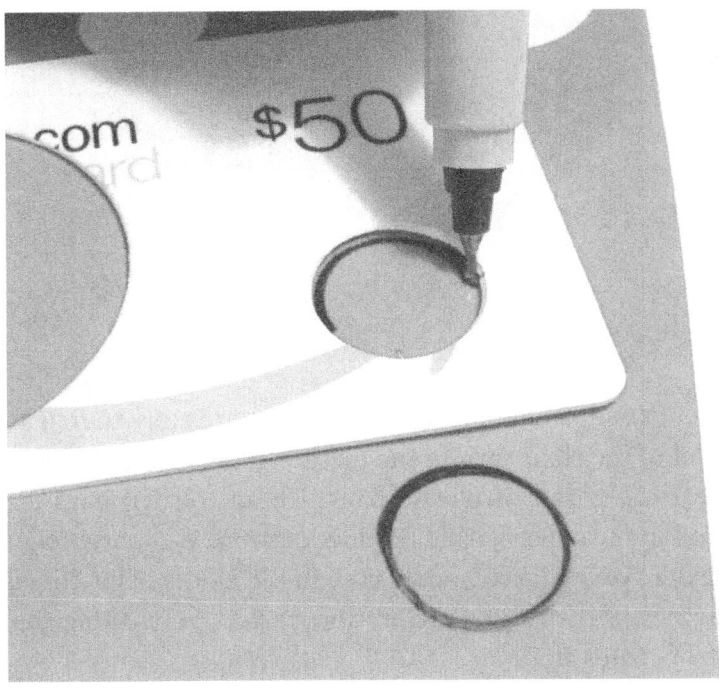

You can even make a hole in a credit card using a punch of the right size, and trace from that as well. Shown here is the credit card method, but the circle template is the best way to go if you don't have a punch of the exact right size. Alternative: you can find some object of the exact size center stop needed, and trace around it. Perhaps you could use a pen top, or lip-gloss cover, or another object such as a piece of PVC plumbing pipe, or metal tubing.

You may have to go to the hardware store to find an object of the right size. In the picture above, I am using the core from a roll of plastic bags. These rolls of bags are found in the supermarket produce section. When the roll is finished, you will see a white core in the holder. This core is exactly 5/8 inch in diameter, and perfect for making that size of center stop.

Trace your circles in fine point marker on your center stop material, and cut them out by hand with a cuticle scissors. This will work fine if you are careful, and is a much better way to go than using a compass cutter to make center stops.

It is in this stage of the process where you will see why I use a set of punches. You will need to make about a dozen of these center stops in different colors, and after doing all that work with a compass cutter, or by hand with a cuticle scissors, you will be a believer in punches.

Your Microscope Hobby

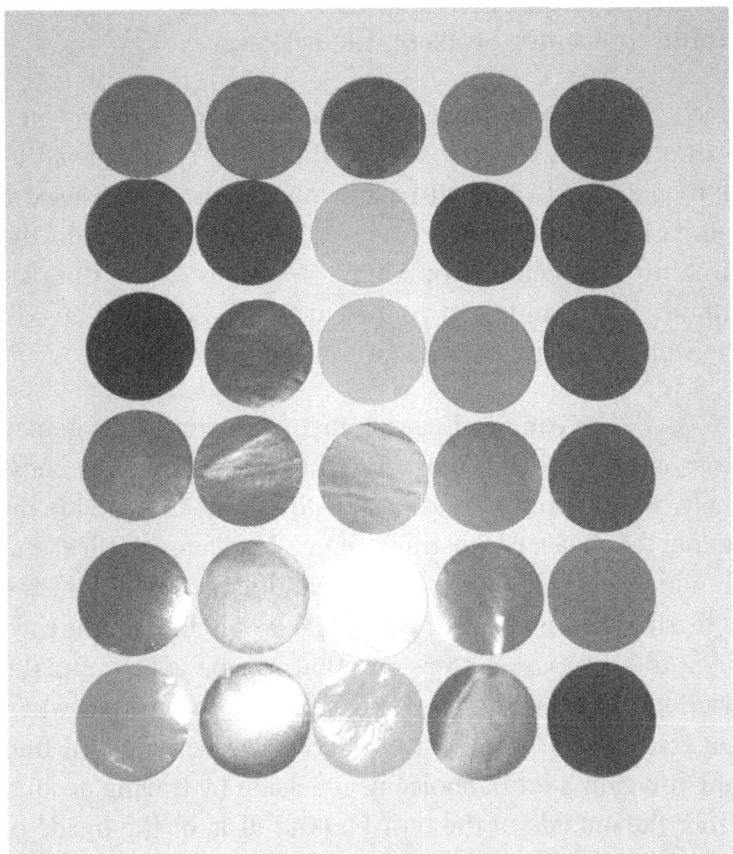

You will wind up with something like the selection in this picture. I purposely left in the glare from the camera flash so you can see the texture of the different center stop materials you can use. Some of it is glossy, some dull, and thicknesses vary.

Mounting Center Stops on Clear Disks

Finally, you have to mount these center stops on a clear substrate, which is a clear disk. You must cut out as many clear disks as you have center stops. If you used a compass cutter to make your test set, you cannot do the same with the center stop filters. You cannot have dimples or holes in the middle of your clear filters, because that will show up under the microscope.

Either you must use a punch, or if you like, you may trace circles using a marker. Your circle template, most likely, will not have holes large enough for the outside diameter of a clear substrate filter. Medicine bottles and plumbing parts should be good candidates to get you the right sized template. You can make a hole in a credit card using the compass cutter- and that can be your template. Trial and error will give you a hole size that works when you trace inside it. If you use a credit card, you will find that you gain a lot of choice in precision by tracing around either the outside of the punched out disk, or the inside of that same circle. There is a range of 1mm to 2mm in this difference. Cutting out the clear plastic with a large sharp scissors on the exact inside of the black line should give you a perfect clear disk of the right diameter.

Once this is done, you carefully mount the colored center stops on the clear disks using a tweezers as you did in your test set. Do this with each center stop, until you have full set of colored center stop filters.

Your set of center stop filters will look something like the set below, depending upon how many different types of materials you were able to gather. Again, I allow a bit of glare in this photo so you can see the clear substrate part of the filters in good contrast. What is difficult to see in the photo are the different shades of color. What looks like blue to the eye, may turn out to be a beautiful Caribbean Sea teal color when viewing your specimen through the microscope.

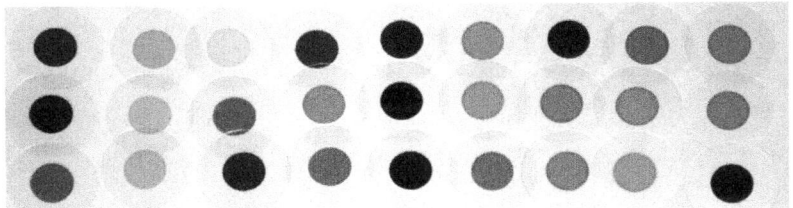

You are done! Your Rheinberg filter set is complete. Here are a few hints and tips on how to get more out of your Rheinberg filter making skills.

Add Colors

Once you have your basic set of filters, go back to the materials sources. Look at the Roscolux web page. Maybe you want to use a peach colored annulus for a pond life specimen of Daphnia. Try and represent what you are looking at under the microscope in the most flattering way possible.

Neutral Density

Make some center stops and annulus rings in neutral density shades. That is a plain gray shade. This material is available from the theatrical gel sources in the Materials chapter. When you have a situation where either annulus or center stop is too bright, use your neutral density annulus or center stop to adjust the brightness.

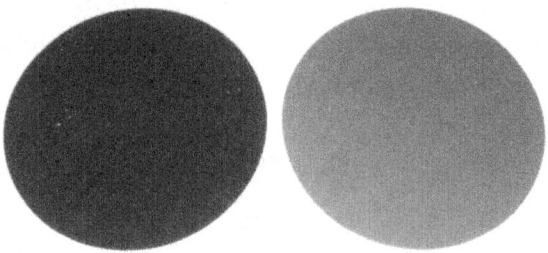

Annulus Mixing

Stack different annulus rings. As you learned in elementary school, red and yellow make orange; blue and yellow make green, and so on. By mixing primary colors, you will get secondary colors. Experiment with the colors you have, and you'll surprise yourself. Under the microscope, a set of stacked annulus rings may give you a really fantastic effect.

This will be best seen in the photo below of two bi-colored rings being overlapped. That's right - two rings overlaid (each split with two colors). This shows you that overlaying rings gives you even more color possibilities. Below are instructions on how to make bi-colored or multi-colored annulus rings.

Your Microscope Hobby

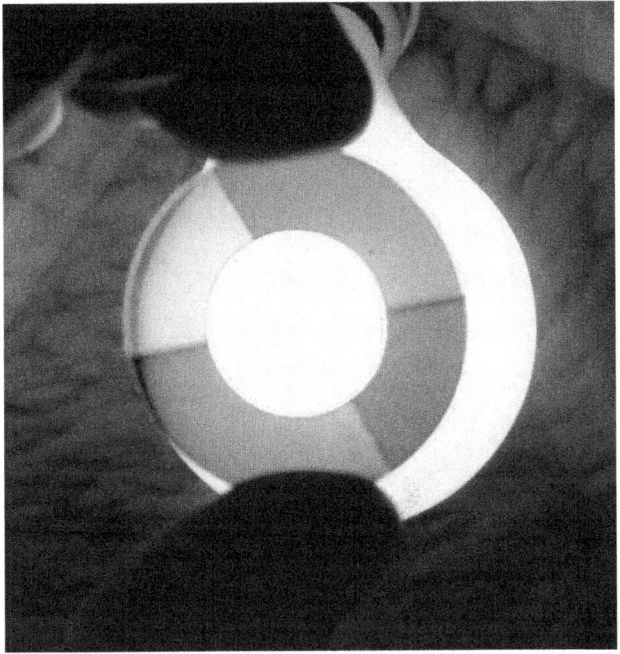

Multi-colored Annulus Rings

Use the Roscolux, Lee Filter, or similar swatch books of theatrical filters. Take two contrasting sample book colors, and lay them side by side. The simplest way is to glue them together using a drop of fletching cement at top and bottom. This type of cement (explained in the Materials chapter) will not peel off the plastic. It dries clear, and quickly. For our purposes, it won't show up under the microscope adversely.

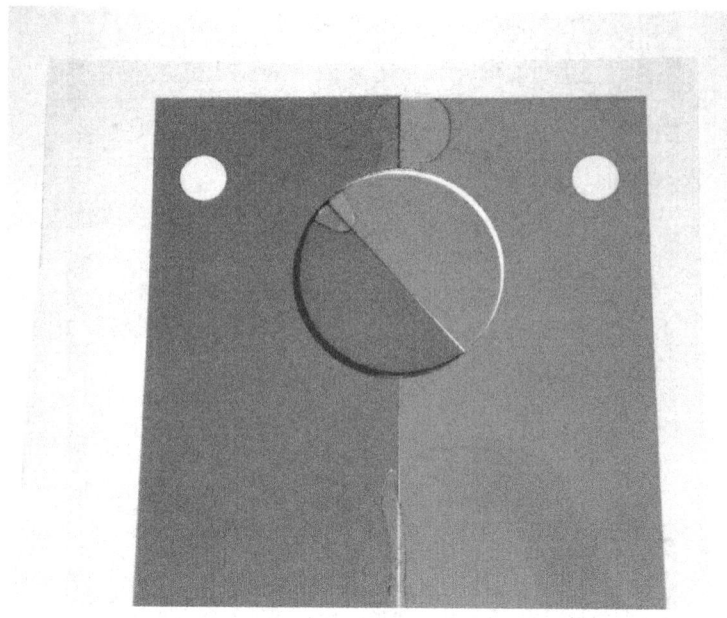

Next, overlay a clear plastic sheet (the kind of stiff plastic you've used for making clear filters). Press it on the the wet cement, flip it over, and let it dry. When dry, punch out your disk, and then punch out your center. In this picture, I've rotated the disk so you can see the single drop of cement holding the two colors in place. Look closely to see the clear substrate material.

Easier still- lay your pieces of sample lighting gel side by side and laminate them. (Lamination instructions below.) Then punch out your annulus rings from this sheet. When you do this with four pieces of gel swatch material you will get a four quadrant annulus ring. See the photos below of how this looks.

Your Microscope Hobby

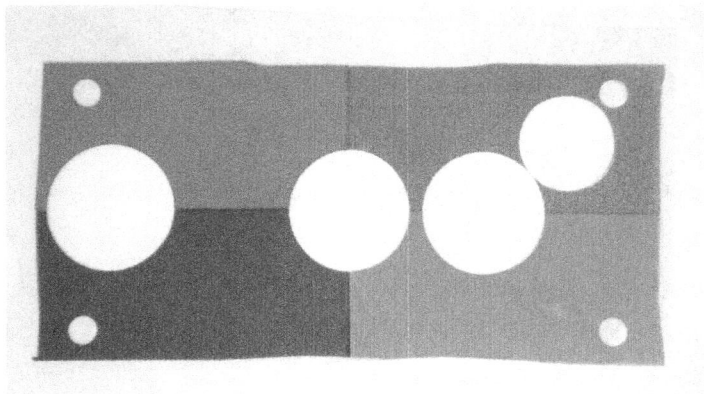

Left side (red top/blue bottom) of above photo shows the laminated sheet with a simple two color annulus hole. This annulus ring will be red and blue. Upper right shows a single color filter hole, green. Center shows a four color, four quadrant annulus hole. And just to the right of it, another two color annulus hole, resulting in an orange and green annulus filter.

If you go to a Staples® store that has a large laminating machine, you can lay out a series of Roscolux filter swatches on a very large sheet. Ask the customer service person operating the machine to do the first pass using a "carrier." The carrier is a white paper sleeve that holds the laminate inside, and protects it from too much heat. The machine must also be set on the slowest motion setting. This will prevent air bubbles in your lamination.

Once that sheet comes out, it will be slightly cloudy, having a frosted look. There will be no air bubbles, and all of your gels will be held perfectly in place. Then have the operator run the sheet through a second time, without the carrier. See below.

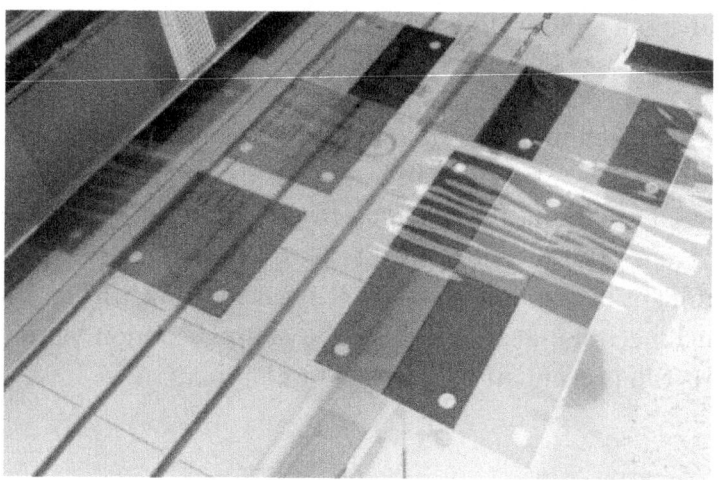

There may be some waves in your laminated sheet, but it will be crystal clear and free of bubbles. Below is how the finished product looks.

You can iron out the waves at home, by placing your lamination between two thin pieces of cardboard and using a regular clothes iron. Be careful not to have the iron setting too hot, and have the steam setting to *off* position.

Above are three unique bi-colored rings. These rings were made with the fletching glue technique, but you can make them using hot lamination as well. And if you overlap differing pairs of bi-colored rings at a 90 degree angle, you'll get even more colors. Against the light, you can see the above three rings in their various combinations with each other.

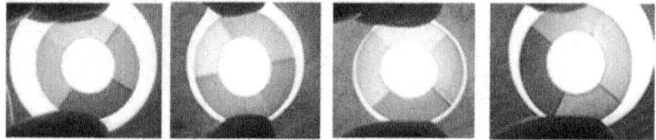

By using lamination, as described above, you can actually cut out four quadrant annulus rings that will look like the above bi-colored rings sets, without even overlaying them.

You can also buy some lamination sheets, and as a DIY technique, simply iron them to complete the hot lamination, using a clothes iron. You must place your lamination sheet on top of a thick piece of cardboard, and over this, you must lay a sheet of card stock. If you let the iron touch the plastic, it will melt the plastic and stick to your clothes iron! So place the lamination sheet under a large 9 in x 12 inch sheet of card stock. Set the iron on medium to start, until you see how your results look. Turn the steam setting to off. As you iron the sheet, gradually work from the sealed end of the sheet to the open end of the sheet, so you slowly move all the air bubbles out. After you punch the filters out from the lamination sheet, you can iron them again to make sure they are flat. Remember to keep them under the card stock as you work.

Finally, you can do very nice hot lamination at home using the 3M Scotch® TL 901 Laminator. It's reasonably priced, and the 3 mil lamination sheets work fine, whether Staples ® brand or 3M Scotch® brand. In the below picture, you can see a set of concentric ring Rheinberg filters that I am making. In order to make rings of this complexity, you really need your own laminator due to the delicate handing required. For colors, you can use the thin

Roscolux or Lee material because it laminates very well and holds up nicely after being laminated. The electrical tape center stops also laminate very well. Therefore, you can make complete Rheinberg filters in one piece, by laminating. This machine is now my favorite tool for making complex filters, and lamination gives you the flexibility of using thousands of colors. By laminating, you are no longer limited to the selection found in heavier Duralar® material.

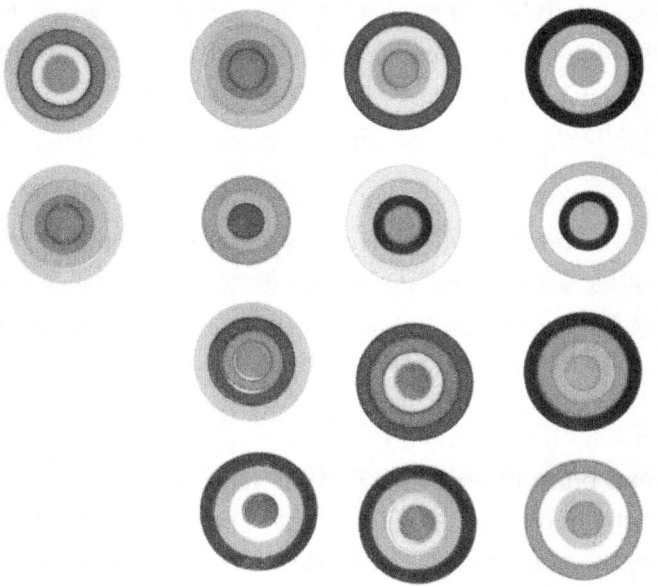

Now, from the above photo, you can see there is another type of Rheinberg filter you can make using the lamination process: Concentric Ring Filters. This is not a drawing or illustration - this is a photo of the actual laminated sheet with the finished filters, prior to punch out.

Multi-colored Concentric Annulus Rings

To make concentric ring filters, start with about fifteen disks in a variety of colors punched to the outer diameter you need. Since you will be laminating these, you can choose from the very wide assortment of colors in the Roscolux or Lee filters. After you have your single color disks made, use a slightly smaller punch to cut out holes in these disks, and that gives you another set of smaller disks (in those same colors). Repeat this process using yet a smaller size punch on your smaller disks, and toss away the remaining centers pieces (or save them. You should now have two sets of rings; the smaller perfectly able to fit into the larger. Lay everything on a plain white sheet of paper or card stock.

Here is where you get creative. Gently place the smaller set of rings inside the larger rings, using complementary color combinations or rainbow order combinations. Complementary combinations would be red and green, orange and blue, and violet and yellow. Rainbow combinations follow rainbow order of red, orange, yellow, green blue, indigo, violet. After recombining according to spectrum or complementary colors, you will have the below arrangement. Below photo shows result of steps one, two, three combined.

Your Microscope Hobby

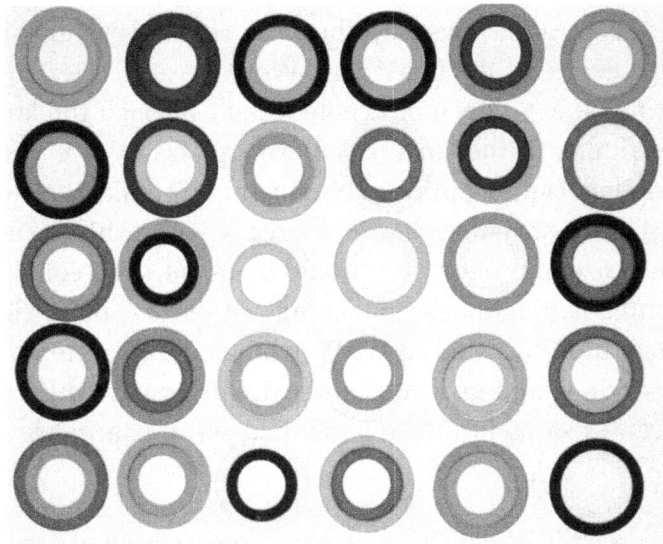

Keep in mind that you can even add another inner ring, using the same technique. Below photo shows where three concentric rings have been made.

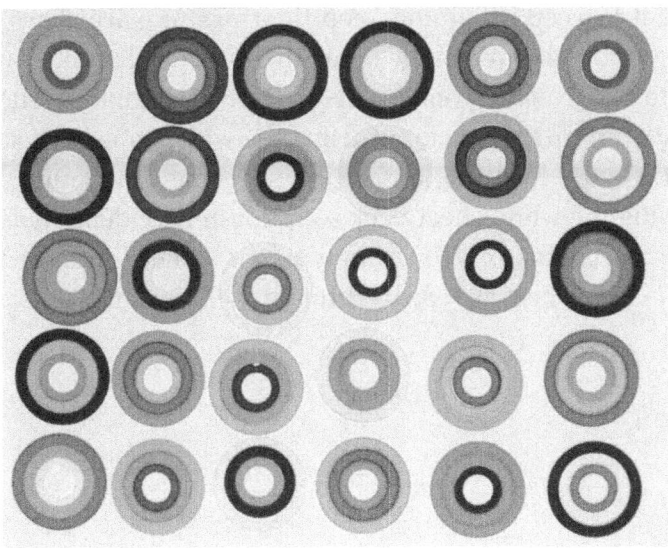

You can always skip the very last step of punching out the center of the inner ring. Why? Because after lamination, you can punch out centers right through the laminated filters, in the correct center stop size. Above photo shows clear centers, prior to lamination, by punching out a hole in the last center section. I already knew with certainty what size my center stop would be. Again- you could do this after lamination by punching out clear centers right through the laminated sheet. Punching out the center after lamination would give you the flexibility to have different size center stops for a pre-laminated set of concentric filters.

At this stage, you now have your combined assembled filter rings ready on a sheet of paper. Open up a new sheet of laminating material, and carefully, using a tweezers, transfer each ring set into the inside of the lamination sheet. Very slowly close the sheet and gently run your hand over it to keep it flat and keep the rings in place. Turn on your laminator. When the laminator is hot, using that same white paper, slide your paper under the lamination sheet. Gently lift up the assembly, using the white paper as a support. Gently feed the lamination sheet into the laminator holding the white sheet back so it doesn't accidentally feed into the laminator. You simply use it to help support the lamination sheet as it feeds through the laminator.

Your Microscope Hobby

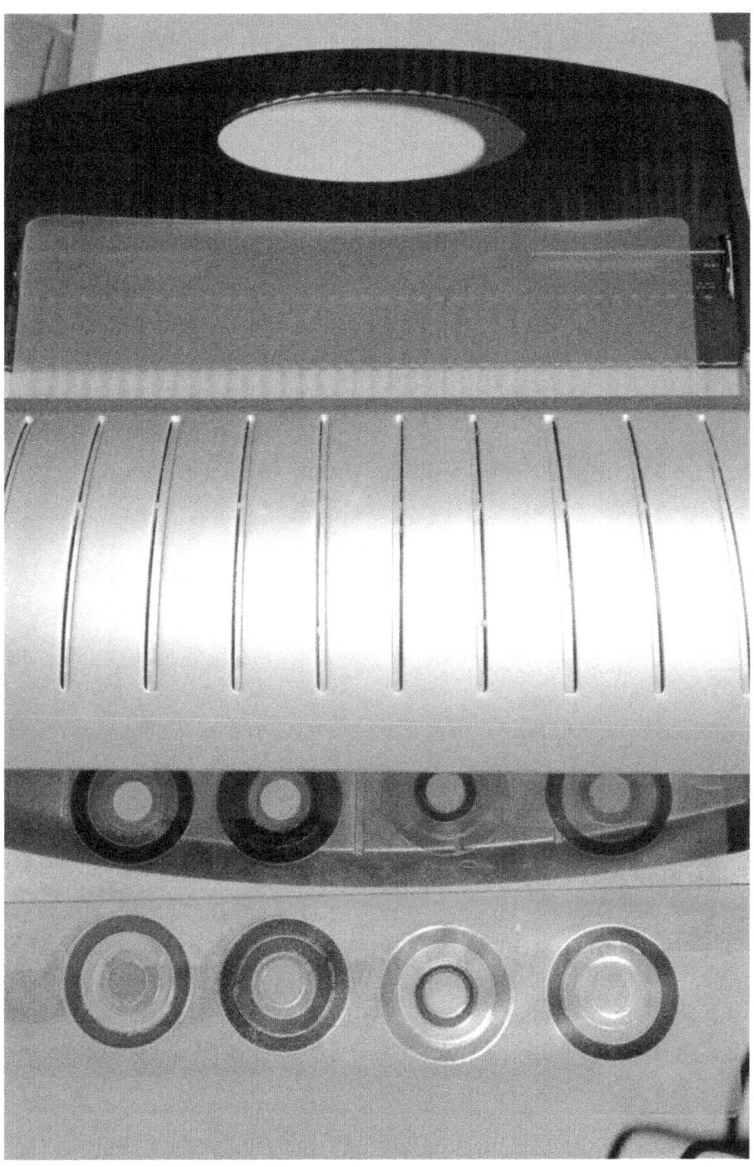

You will be amazed at the result. Now, all you only have to punch out your disks using the desired size of outer diameter punch. If needed, punch out the centers of these disks to the exact size of the center stop you will be using. In the above photo, however, I have actually completed the filters with center stops made of colored electrical tape. The sticky center stop disks actually help keep each concentric ring filter in place within the lamination sheet. You do not need to do this, but it is definitely nice if you already know what color combination you want.

Below is a clearer photo, where I placed a piece of white paper below the filters just for the picture- so you can see the filters more clearly. The lamination sheet is to be run through the laminator by itself- without paper under it.

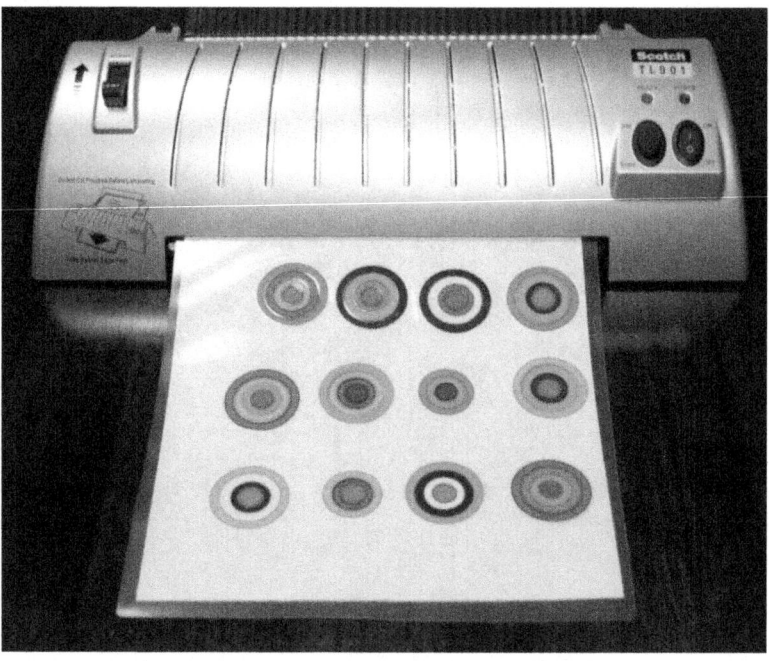

How a Full Set of Rheinberg Filters Looks

It's important to see what a full set looks like. If you put photos of sets for sale on your website or on eBay, you must have a disclaimer that the final set will not be exactly as shown, due to differences in microscope specifications, and subject to availabilities of materials.

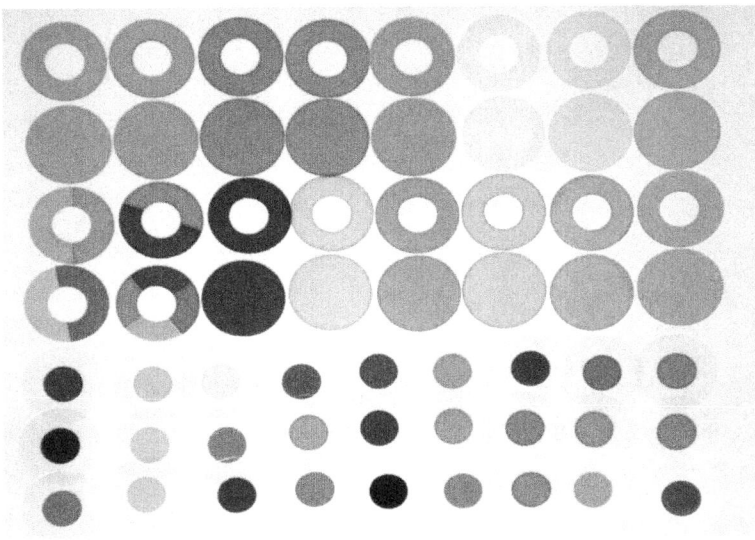

CHAPTER 6

Polarization

You're an expert on punching out filters. Well, all you need now is some polarizing material, and you can enter the world of polarized light microscopy. You need one filter below the condenser, and one above the eyepiece. I have actually cut out small disks as well, which fit over the eyepieces, using rigged up holders from old 35mm film containers. You don't have to get fancy.

Basically, you could even wear polarized sunglasses or 3D specs from the movie theater. By the way- both are good sources of polarized material. I was lucky enough to obtain self-adhesive polarizing sheets from American Science and Surplus, which they carry from time to time. Subscribe to their catalog, so you don't have to keep checking their website. Once you order from them, they will send you their catalog every time they come out with a new one. See the web address above in the Materials chapter or in the Appendix.

Once you have your polarizers ready, rotate the bottom polarizer after you have your specimen in focus. This rotation will reveal wonderful characteristics that you simply will not see in Brightfield. Take these crystals, for example. Polarizers used in this photo reveal different crystal formations within the same specimen that are apparent when rotating the polarizers. Crossed polarization is when the polarizers are turned at 90 degrees to each other. You can see what happens when you hold up two pairs of sunglasses at 90 degrees to each other.

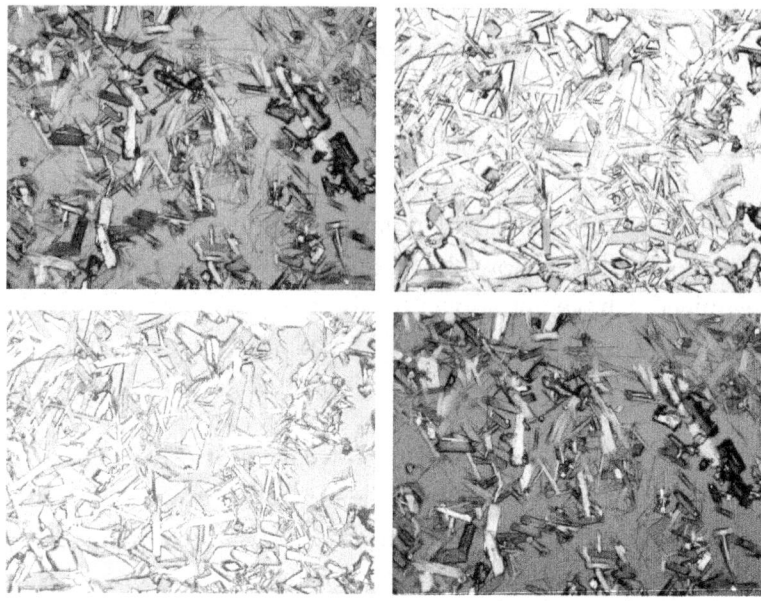

Next, you need an old plastic CD case. These are also called *clamshells*, because of the way they open. Take it apart so you have the clear top only. When you move and rotate this plastic just above the lower polarizer, you will get some unbelievably beautiful colors. This is what they call an interference filter, but to you it is simply a plastic CD case lid. The most astounding colors are obtained when you combine this CD case rotation with rotation of the bottom polarizing filter. My best crystal photos were taken this way, and above, you see a comparison with and without polarization, and when rotating the CD case.

Polarized Rheinberg and Variable Darkfield

You can also make yourself some center stop filters and annulus rings of polarized material. How can you use a polarized center stop with no annulus ring? If you place polarizing filter (or disk) under the condenser, and rotate it, you can actually control the brightness or darkness of the background only. This is Variable Darkfield, a new concept. The specimen brightness should be unaffected by polarization.

Can you use a polarized annulus ring with colored center stop? Why not? If you place another polarizing disk under the condenser, and rotate it, you can actually control the brightness, darkness, and the polarization of the specimen only. The background is unaffected by polarization, and remains the color of your choice (the center stop color). Polarized Rheinberg is another new technique for microscope contrast.

If you like polarization, check out Spike Walker's work with crossed polarization and Rheinberg illumination. "Astounding" would be an understatement. Phenomenal is closer to how I would describe his work. It's all at the link below:

> http://www.microscopy-uk.org.uk/mag/indexmag.html?http://www.microscopy-uk.org.uk/mag/artjun05/swgallery2.html

By the way—Micscape Magazine is wonderful free E-magazine, which has great articles on all things microscopy. Spend some time exploring this excellent resource. The main website is:

> http://www.microscopy-uk.org.uk/

> You'll be amazed at all the great stuff there.

CHAPTER 7

Oblique or DIY/DIC

The expression DIY/DIC (Do It Yourself – Differential Interference Contrast) came from an article by Wim van Egmond published in the on-line Micscape Magazine, which you can read here:

>http://www.microscopy-uk.org.uk/mag/artnov02/diydic.html

Okay, it's not really DIC. It is actually "oblique lighting." It's been around forever, and it is the poor man's DIC, but don't underrate it. It still does a great job of showing relief on the microscopic level. It's the difference between a flat cave painting and a sculptured bas relief on a Greek temple. And when it comes to clear micro-organisms, that's huge. Both DIC and oblique lighting provide contrast, and contrast shows how something looks in reality.

Wim van Egmond's filter looks like this:

And he suggests adding a colored Rheinberg center stop overlay, like this:

Although Wim calls it DIY/DIC, we will simply call it oblique lighting. The light beam does not go straight up through the condenser, but is redirected at an angle, obliquely. The angle creates a slight shadow effect. Get up really early in the morning and look at a landscape at sunrise. Why does it look so beautiful? Because the light is low, near the horizon, and as it streaks across the vista, it casts shadows and provides all the detail in relief. Take a look at this photo, taken early one morning.

Something about it is beautiful. It's just a landscape snapshot, but why is it such a nice photo? Why do you feel like you are there? It's the lighting. It is early morning light from the side. Look at any example of great photography, and you will see that the hour of day is all important in getting a fantastic shot. You don't need an orange sunset or snow-capped peaks to capture a great landscape photo. You just need to get up really early, and take the shot at a 90 degree angle from the sun – with the sun streaking its light from the side.

Now you are ready for oblique lighting in photomicrography. Some of the old condensers were actually able to move sideways to provide oblique lighting. Modern condensers often have a swing in lens underneath. This lens is used for low powered objectives. You can swing this lens part of the way in, and thus bend the light, blocking some of it. You will get oblique lighting. And finally, you can make special filters that block some of the light, and create oblique lighting. Take a look at this picture of a "cast" of timber wolf hair using oblique lighting. The only way to see the detail in the surface of the clear cast is by using oblique lighting. What is a cast?

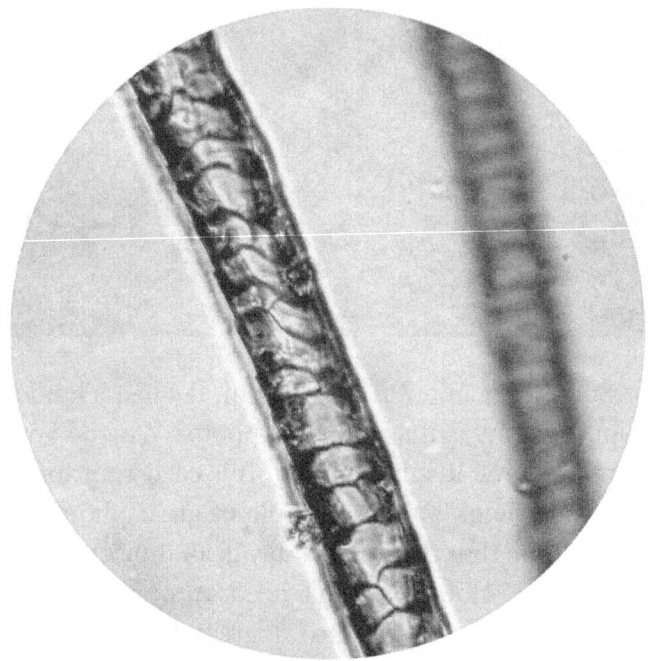

A cast is made by painting over a specimen with clear nail polish, letting it dry, and then peeling away the specimen from the coating. What remains on the slide is an exact impression of the surface of the specimen. This is nice, because you can see the texture of a specimen, which would have been hard to see when looking at the original. In the case of animal hair, you can actually determine the type of animal by looking at the texture of a single hair. If you want to try this, it is a matter of trial and error. Simply paint your slide, and press your leaf or hair into the clear nail polish. Allow it to dry, sometimes it's best if not completely dry. Then peel off the specimen and look under the microscope. To see the texture, you must light it from the side with a narrow beam of light.

Making filters for oblique lighting is an experimental affair. There are many ways you can block and redirect the light path, and what works for your particular microscope is literally a shot in the dark. So, I can share with you what has been tried, what has worked for some people, some of my own creations, and the methods of making these filters called DIY/DIC filters.

First, take a look at these diagrams, and then I'll explain how to make the filters that match. Make a semicircle of black to start.

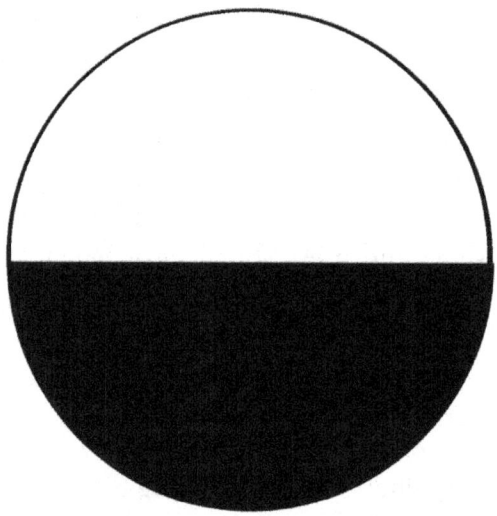

Then make another so you have two filters, half of the area solid black on each.

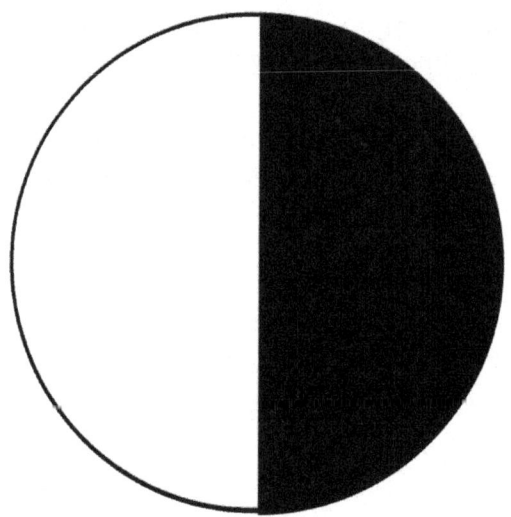

Overlay at right angles, and you can create a 3/4 black filter.

Your Microscope Hobby

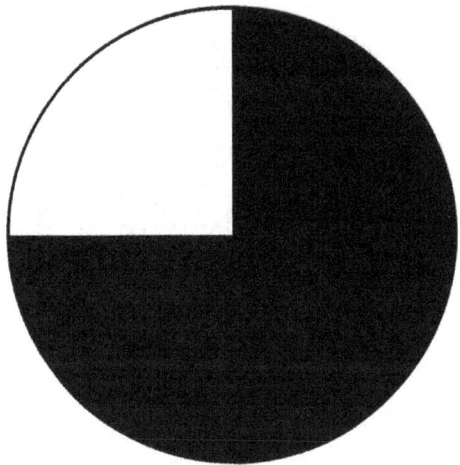

Rotate the two semi black filters varying by any amount, and you have an adjustable pie slice of light.

Next, you can overlay a black Darkfield stop filter over the previous pair of filters, to further reduce the amount of light. Center stop is shown in blue for illustration, however it would be black in reality.

Next you can make a series of crescents in graduated phases.

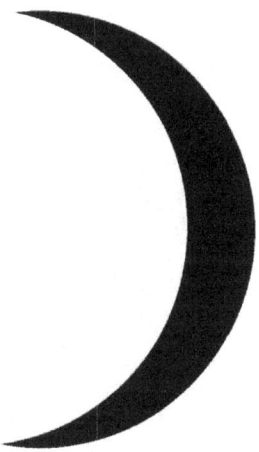

These can even be cut from old credit cards, as you only need the opaque portion.

Oblique lighting tends to work best when you have only a tiny bite out of an opaque background, as shown below.

And finally, you can come up with a variety of lines, shapes and patterns, and combinations of those.

The goal is to find a combination of filters that will give you a 3D relief effect with a specimen that you would not be able to see in Brightfield.

Your Microscope Hobby

With the correct oblique filter, you'll achieve astounding depth, as in this photo of a single diatom using crossed polarizers to add some color.

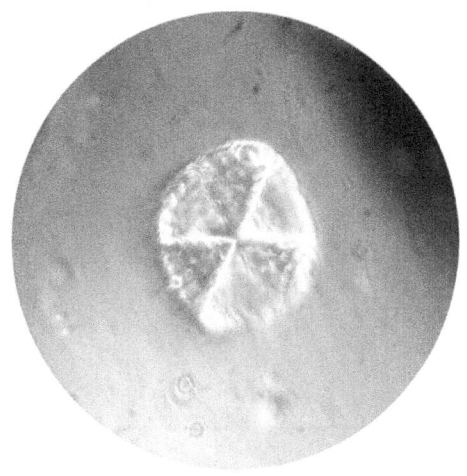

See the below photo of diatoms in relief.

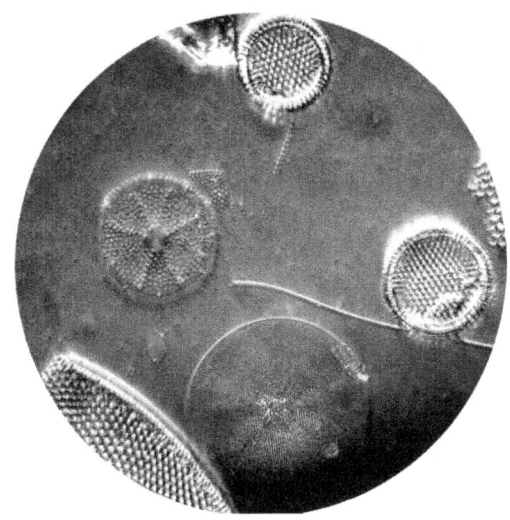

See the below diagram for an idea of how the light moves up through the condenser and shoots across the specimen sideways to provide the oblique effect.

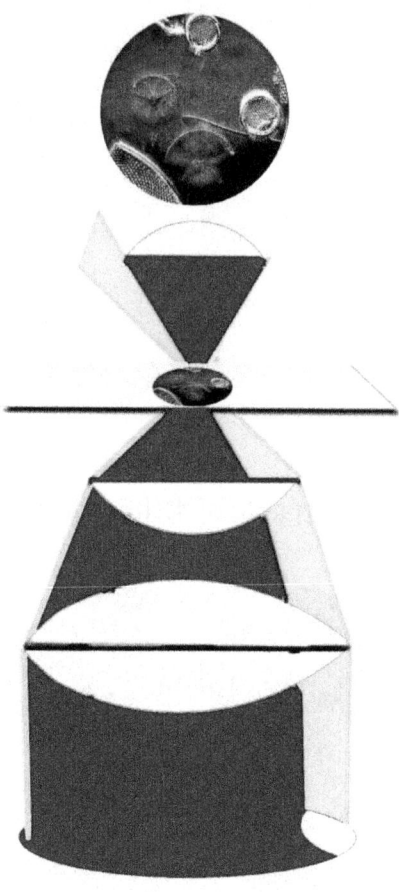

Your Microscope Hobby

The above light path illustration would be using the solid oblique filter with a slight bite out of one edge.

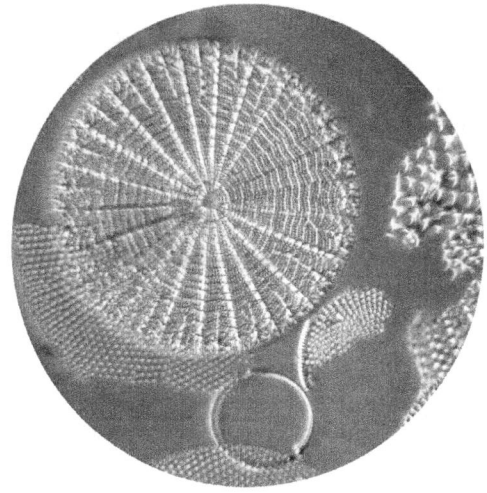

And if you combine a Rheinberg center stop with an oblique filter, you could get a colored background as in the above green picture.

The nice thing about making oblique filters is that you can use cut up pieces of cardboard or plastic credit cards. The card keys from hotels you've checked out of, and those used gift cards really are perfect. They are durable, easy to cut, and totally opaque.

You should even be able to cut out filters like the one designed by Wim van Egmond in the article cited above. It will take some work, and I have done a number of those, but you will have to discover the best technique that works with your tool set.

The best material with which to make a good variety of oblique filters is black self-adhesive vinyl. You can set up geometric patterns on a clear sheet of plastic. You can make strips in a variety of widths.

Then, punch out your disks using different sections of the black and clear. Here you see in each portion of the above picture, how you can create many different ways of combining black and clear on a disk. These disks will later be combined with each other to allow just the right amount of light in the exact right spot for your condenser. By experimentation, you should be able to get some basic oblique lighting effects.

Eventually, you will have a stockpile of shapes, and you will continually be experimenting with various combinations. When you hit upon something that gives you the oblique DIY/DIC relief effect, try to make a single filter that incorporates all of the qualities of that filter combination.

Shown below is just part of a set of DIY-DIC filters I made which includes some negative versions of the previously shown oblique filter diagrams, to give you some ideas. For each design, you can also make its negative opposite. Making these filters is really a matter of personal preference and ingenuity.

Below is another good link to a site that shows you this technique in action.

http://www.microscopy-uk.org.uk/mag/indexmag.html?http://www.microscopy-uk.org.uk/mag/artfeb00/pjoblique.html

Finally, if you are selling filters, oblique DIY/DIC filters are a nice bonus you can include for free if you wish. Of course, you can also charge for sets of DIY/DIC filters, however you really can't guarantee that they will work. So you probably want to charge only a small amount to make these, like $25.00, and add the caveat that they are experimental and not guaranteed to work. They are relatively easy to make, and actually fun to create, so for the small charge, you can be persuaded to make an occasional set.

How a Typical Set of Oblique Filters Looks

As mentioned earlier, if you put photos of sets for sale on your website or on eBay, be sure to have a disclaimer that the final set will not be exactly as shown, due to differences in microscope specifications and subject to availabilities of materials.

Your Microscope Hobby

CHAPTER 8

Cases

You can get this great filter case in my Amazon store, or with some searching at a major supermarket or drug store chain.

These cases, intended for vitamins and prescription pills, are made in the full rainbow spectrum: red, orange, yellow, green, blue, and violet, as well as clear. The case comes with an extra top, so you can even split it into two stacks. If you purchase two of these cases, you'd have a holder for all of your center stops, and a holder for all of your solid filters and annulus rings. Organize your filters any way you like. Don't forget to make disks of craft foam for the bottom of each section. You want to protect the filters from scratching.

Above is how a filter case looks in all its glory when filled with Rheinberg Filters.

The major craft stores have similar tower cases in clear plastic. They are to be found in the bead making section. People who work with beads use these stackable cases for their various colors of beads. Be sure to scour the art supply stores and discount stores as well. They always have nice cases and boxes for your microscope equipment.

Your Microscope Hobby

Another good type of case for individual filters is a ladies cosmetics make-up compact. You can pop out the metal makeup tray with a fine screwdriver, and you've got a great case for one or two special filters. These come in different sizes, so plan to raid your wife's cosmetics box. Don't forget that prescription bottles also make good cases for a vertical stack of filters, or filter blanks.

Objective Lens Holder

Here's a quick project you can do to make a holder for your objective lenses. Occasionally, you might find yourself knocking over an eyepiece or an objective lens that sits out on your work surface. If an eyepiece or objective rolls off the desk and hits the floor, it can be a very costly accident.

Simply take a 1 x 2 inch strip of lumber and cut it into short sections. Drill holes approximately 1 inch in diameter along the sections. Sand and coat with varnish, and you have a nice holder for your objectives. You can even mount these in a larger box, and you will have a great case as well.

How to Build a Filter Box

This box will be perfect for glass or plastic filters, and act as a great way to have all of your important filters at your fingertips. At the same time, when you are not using your filters, the case acts as protection from dust and damage.

Your Microscope Hobby

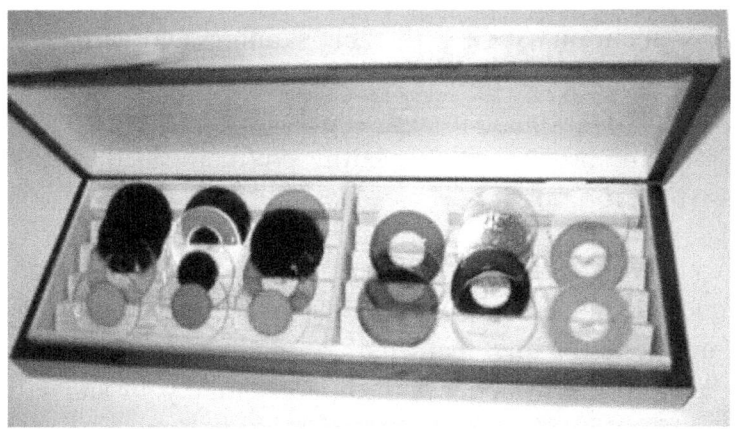

Start by going to the crafts store, and picking out a long flat wooden case. You'll find a nice variety of these cases in the wood projects section. Below picture is straight, however you are experiencing an optical illusion if you think it is crooked on your screen.

Plan on sanding, possibly staining, and varnishing the finished case. Above is the case I selected. I bought two, so I could make an extra one to sell on eBay. It paid for the cost of the project with a healthy profit. I sold that extra case for $100.00.

Next, you will need some wooden slats. Buy this wood in the molding section of your favorite hardware store chain. Get an 8 foot length or so of plain, flat molding in a width that is about ½ inch larger than your standard filter size. If you are using a 32mm filter (1 ¼ inches), your wood slat molding should be about 2 inches wide. The thickness should be about ½ to ¾ inch.

I won't give you exact numbers here, because people have a different filter sizes, and will choose a different sized case. So look at the following pictures and the text description to understand what you need to do.

With a pencil, draw a line down the entire length of the molding, dividing the molding in half, lengthwise. Then, mark off the molding in section lengths that will fit inside the box you purchased. Finally, mark the molding with small markings representing the center of each filter. Be sure to leave enough space between the filter markings to allow about ¼ inch between each filter. It should look like the above picture. Leave yourself about two feet of blank unmarked wood at one end. You will need it later.

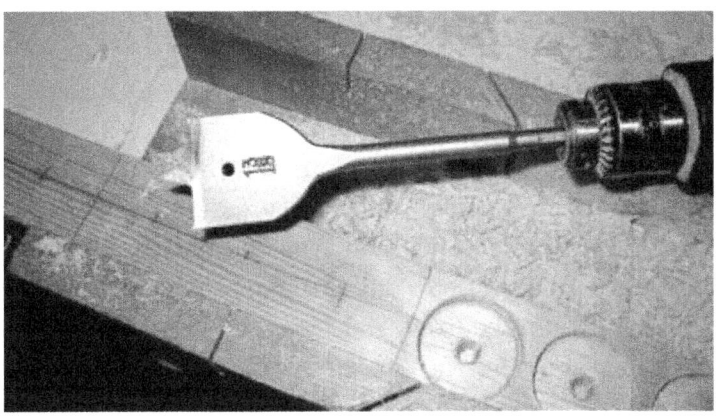

Now, you will need an auger bit that is slightly larger than the size of your filter. The auger bit must make a depression that will allow your filter to fit not too loosely, and not too tightly. If you have a router, this is even better. I used a size 13/18 inch auger bit for my 32 mm filters. Drill out depressions at your filter markings, along the entire length of the molding, as deep as you can go, without coming through the other size of the wood. Again, you should leave a couple of feet of wood untouched.

You can see from the photo that the point of the bit leaves a hole in the center of the depression. That will not matter, however, because you will be splitting the wood lengthwise. After you have all of the depressions done, you can then slice up the wood slat into sections at the vertical markings you made. You can these marks in the in the pictures.

After you have the long wood slat cut into sections, you can proceed to saw them in half lengthwise. I used a hand-held jigsaw to do this.

Your Microscope Hobby

You will wind up with a stack of sections, each with half round circular indentations cut into the wood. If you stack them together, you can test how your filters will fit. Notice that I show a couple of pieces of un-drilled wood. We will use these to finish off the front part of the filter tray.

Now, use wood glue to stagger and cement the filter trays. You must use clamps to hold the wood in place while it dries. Stagger the trays so a filter can slide into a circular groove and remain in place without toppling over. You'll have to test this by hand to get it right. Here is how it should look.

You will have to work through the glue process by setting up two trays at a time, and clamping both ends. In the picture here, you see that I have already done two sets, and I am clamping the two sets together. Depending upon the size of your filter box, you will have to adjust how many trays you assemble, and how long these trays are.

After the filter trays are assembled, you will need to cap off the front tray with an un-drilled flat piece of your wood slat. Otherwise, those filters would simply fall out.

Then, flip the assembly over, and sand flat the bottom. This is optional, and will depend upon the interior height of your filter box. I selected a fairly flat filter box that would not close due to the height of my filter tray with filters. So I had to remove some height, and I did this by sanding flat the bottom of the filter trays. Here are a couple of different views of that result to illustrate.

You only need to sand, stain, and varnish the components. Then assemble them. Below is how my finished filter box looked on my desk with my microscope set up.

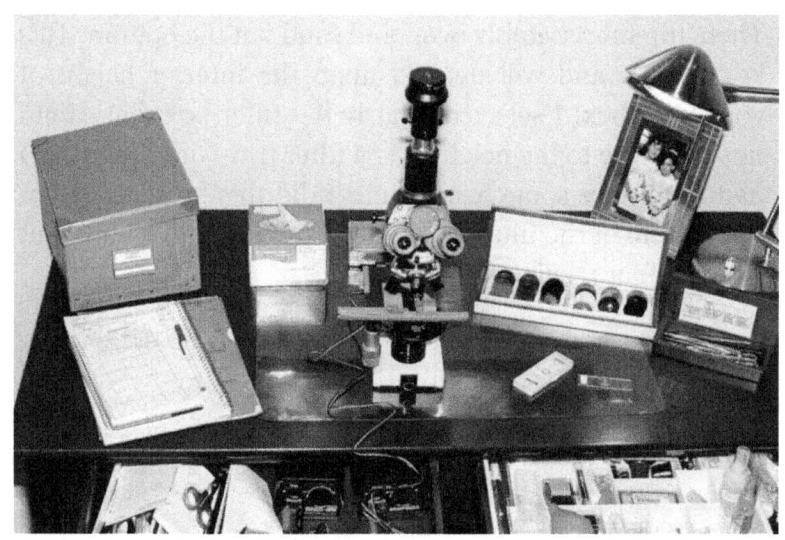

CHAPTER 9

Selling Filters for Fun and Profit

Here is the kind of feedback you can expect to look forward to.

> *"As a newbie to the field of microscopy, I found Mike Shaw online in searching for information on filters for my new 'scope. He has been just great, working closely with me to build a full set of filters to fit exactly, and providing great guidance along the way. You will find him most patient and very knowledgeable, as well as fair in his pricing and very responsive to questions. A great resource to amateur microscopists everywhere!*
>
> *—Pam R"*

If you just do just a few of the following things, you'll get plenty of filter inquiries. That's all you want. Inquiries. Once you have inquiries, you can convert those into sales by presenting your pitch. This chapter covers marketing techniques, how to respond and pitch, how to fulfill orders, and how to follow up.

Sell Filters on Your Website

The easiest way to sell filters is to create a web page and let the inquiries come in. This has always been my best source for steady filter sales. You don't need a special domain for this, and any webpage or site you have is fine. The search engines look for content, and if you have "Rheinberg filters" on your site, you will get hits. I used my AOL webpage for years, and later used my tardigrades.us website, which has a separate filters page. In fact, if you use WordPress, you can simply add a Rheinberg filter page to your existing WordPress site, as I did.

Only recently did I set up a website specifically for Rheinberg filters, simply because it became so affordable with cheap domains. You can look at it, if you want to do something similar.

http://www.rheinberg-filters.com/

If you want an inexpensive domain, I recommended GoDaddy earlier. It's a bit of work to set it up, but you can get free hosting and it includes email, and it ends up being a very low cost annual expense. I did that myself in the beginning. If you purchase your own hosting, it's only slightly more expensive, and then you can put many different websites on your single hosting account.

If this is confusing, here's a quick and pain free lesson on websites.

A "domain" is a website name. www.mikeshawtoday.com is a domain. A domain can also be a .com, .net, .org, .biz, .info. Therefore, domains, are the names of your websites—with many types of endings. You can buy domains at a very reasonable price. The .com domains are more expensive. If you go with ".biz" or ".net" or ".us," like I did for my tardigrade site, it is very cheap. Here are some domain names that I use:

www.tardigrade.us represents tardigrade information in the USA.

www.mikeshawtoday.com is a personal web site

The above are examples of domains. How do these domain names exist on the internet? They are *hosted*.

Hosting is that magical internet connection that lets your domain appear on the World Wide Web (www). Here are examples of hosting.

GoDaddy – offers free hosting, as well as paid hosting.

Your own computer can host your own domain. That's free, but you have to configure it, and leave your computer on all the time. Not recommended for the beginner.

Other hosting sites like HostGator or Bravenet are good too. Bravenet does have lots of nice tools to help you with domain building.

Google the phrase "domain hosting," if you want to see all that is out there.

In summary: You buy a domain name, and then you need hosting to make it exist on the internet. Hosting can be free when you buy a domain.

What I like about GoDaddy for newbies is that they have free hosting, free photo albums, free email (sometimes), all with your own domain name. You can purchase website names (domains) for around $10.00 per year. Some have been as low as $2.00 per year with all the free stuff! Best of all they have great 24-7 phone support, and they walk you through the set up.

http://www.godaddy.com/default.aspx?isc=IAPtn0100

If you use the above link, I will get an affiliate commission, so I'm putting that out there in the interest of disclosure, however it will not affect your price.

Once you have your domain and hosting, you will need to build a web page. The best way to do this for free is to learn to use WordPress. I'll explain that in a minute.

But first, a word about learning new skills: Don't pay anyone to do anything for you when building your website. Bite the bullet and learn some new skills. It will take time, but the payoff is great personal satisfaction, and it may even bring you extra income.

Building Your Website

Using WordPress. WordPress is nothing more than a web building add-on that you attach to your website. You simply pick out a template that you like. A template is a style or a look you want your website to have. Once you have WordPress you have the capability for web pages with a professional look, plus many features like interactive blogging, and the ability to build a mailing list, and so on. It is all very easy. There are thousands of free templates to choose from whether you want your site to display your photography, sell your Rheinberg filters, to promote your catering business, to be a newsletter, or be the family scrapbook.

Someone may recommend to you a site called WIX. I've tried it, and it is clunky in my opinion. It is not nearly as easy to use as WordPress, and you are always stuck with the WIX logos, unless you pay a monthly fee. Go with WordPress because there are built in spam blockers which are free, and are very valuable. I know html code and for years I coded and built my own web pages. Now, all my sites use free WordPress templates.

If your site is hosted on GoDaddy, you simply push a button that says "Install WordPress." That's what I recommend. If your site is not hosted by GoDaddy, you will have to go to the WordPress site, and install it from there. That's easy, too.

Select your template, and start adding content to your pages. Have fun with it. Leave one page for Rheinberg filters, and make the rest of your site about microscopy or about your other hobbies. Be sure to check out my website www.tardigrade.us and see how I did this. Above is a link to my personal website and will give you some more ideas.

Your Rheinberg filter website or webpage should tell people three things:

Tell about Rheinberg Filters

Tell a bit about Rheinberg filters, and show picture of what they do. Show some examples of microscope photos you've actually taken with your own filters, and pictures of the completed filters in all their glory. Do not show how you make them, or anything about the materials you use, or the tools. These are what we call trade secrets. You have paid the price of this book, and invested your valuable time reading, and your precious effort learning how to do it. Why expose all this information and let someone say "Oh, I can do that." That just downgrades the price you can charge for your filters. You don't want to quote $100 for a set of filters, and have someone say, "I'm not paying that much money for some electrical tape stuck on a piece of plastic from report cover."

Just as a magician does not reveal his secrets because it will spoil the show, don't allow your hard work and practice at your craft to be downgraded. What makes magic great are not the secrets, but the performance and the art. What makes your filters great is the awesome beauty they provide to the person looking through the microscope, not the materials or tools you use to make them.

Tell about What You Are Selling

You need a statement saying that you make and sell Rheinberg filters. You can show a price, but I recommend against that. Allow some wiggle room later. Another reason not to show a price is because if you are selling to someone in Alaska or Russia, your postage costs will be higher, and if you have to send more than one sample set, you need to cover that expense. If you have no price showing on your website, this forces your prospect to send you an email. You also get to build up a mailing list of prospects. And when you have an email address, you can go back in a month if they haven't ordered, and ask if they are still interested. You can say you have some new colors in stock. You don't want people looking at your site, thinking the price is too high, and leaving the site. You get nothing from that but web hits. When prospects send you an email asking the price- you can explain the process of the test kit, say that you make the filters specifically for their scope, what colors they will receive, and so on. The email is where you make your pitch, not on the website. Prices on a website should only be for ready-made items that can be shipped immediately.

Next, mention a bit about what is included in the set you are selling. Don't be too specific, because as you add more colors, your set size will change. If you are making a set of filters for someone who needs a 20 mm center stop, you cannot use the narrow rolls of colored tape. He may only get six center stop colors in his set. Show some filters, and say a set "typically" has this many filters. Say that each set is custom made for the buyer's microscope.

Provide Your Contact Information

Your contact info should not include your phone number or address. Show only your Email address. You do not want to put your phone number on the internet, because you'll get lots of solicitations and prank calls. A real filter prospect may leave a message that you cannot understand. You don't want your actual physical address on the internet either, because you will get junk mail, or worse, visitors. Even if you have a PO Box, you could get junk mail or things you do not want. You don't want someone sending you a check for a set of filters, when you haven't even gone through the test process. Your email address is fine, and that's all I've ever used.

Once you have set up a website for your Rheinberg filters, you can begin selling and promoting your filters. To receive payment, you simply invoice the buyer through PayPal. It's that simple. Almost everyone has a PayPal account now.

Do you have a specialty in microscopy, or in another hobby? Perhaps you would like to write an e-book, or sell hand tied fishing flies, or crafts, or art. In that case, all you will need is a website, an "Add To Cart" button and PayPal.

Would you like to have a real "BUY NOW" button on your website? You can have a button that says "Add To Cart," or "Buy Now." PayPal now has a function that allows you to create payment buttons and insert them on your website. Clicking the "Buy Now" button takes the buyer right to PayPal and your invoice. That is only for items ready to ship. Don't use this button for your custom filters.

Sell Filters on eBay

There are people who afraid of selling on eBay and Amazon because of the hassle factor. What is this hassle? You have to learn some new skills. Yeah, you'll have to learn how to navigate in your eBay or Amazon account, and how to set up PayPal, and get paid by Amazon. Well, if you want extra money, you do have to do some work. Right?

I have made thousands of dollars selling on eBay and Amazon, not even trying to make money. It was addictive. Let me tell you what happened.

It started out as me wanting to get rid of a lot of stuff I had accumulated over the years. I had this old metal lunchbox from elementary school, with the *Seawolf* submarine theme. I just didn't have the space to store a lot of stuff, so I listed it on eBay. It sold for around $50.00, plus shipping, and I was hooked. I especially liked the fact that I wasn't simply parting with this collectable item. I was giving it a new home, where it would be appreciated. I was not letting it go for pennies like you have to do at a yard sale. People who really wanted it were bidding on it.

I did this a few more times with an old camping cook kit, my metal detector, my Planet of the Apes movie trading cards (from the original Charlton Heston movie), and a few more mementos. My wife was downwind of the smell of extra money coming in and it was like a shark getting a whiff of blood.

She was tired of polishing all the silver we kept in the china cabinet, and we listed it on eBay. Again, in came the money, and out to a good home went the items, resulting in less work to do around the house. Now we got serious about cleaning out the basement, attic, and all closets.

We had hundreds of books that I wanted to get rid of, and I was able to list about 75% of them on Amazon just by typing in the ISBN number, and clicking a button in Amazon's "Sell Mine" section. Not a week went by that we didn't have two or three book sales. Was it work? Yes. I had to learn how to pack efficiently, and how to ship cheaply, and how to buy postage through eBay and Amazon. I scrounged boxes and cardboard from the supermarket, and bought rolls of packing tape. This is one of the for-real ways to earn extra income at home. Later when my daughter was in college, we sold her used textbooks in Amazon's used college textbook section. It was a huge savings on college expenses right there.

It was then that my wife and I had a sudden realization. We could actually buy stuff and then sell it to make money. What a concept.

We started going to church rummage sales, library used book sales, and neighborhood yard sales. We already had a pretty good idea of what would actually sell on Amazon and eBay. We bought bags of books for 25 cents a bag and sold the individual books at $15.00 to $20.00 for a nice "coffee table" book in excellent condition. Back then, we'd find some guy at a yard sale selling video tapes for $1.00 per tape. We'd buy his whole table full of 100 tapes for like fifty bucks, and then sell a particular movie genre, or actor's collection set of tapes on eBay, in lot sizes of 5 or

10 tapes to an auction. We'd find Depression Glass bowls at yard sales for 50 cents and sell them for $10.00. Check out our seller volume and rating on eBay where we are shown as seller "theshawsrus."

By learning a new skill, you earn additional income to supplement what you are making at your regular job, or your stay at home partner can actively run a fun business and bring in additional cash. Did I say cash? Yes. I won't go into the tax benefits here, but you get the drift. Check out eBay and Amazon's policies to see how high the threshold is on sales before you have to pay tax. So I urge you to learn some new skills, challenge yourself, and make some extra dough.

After you learn how to make filters, and you are ready to sell a set, you can copy my ad, which you'll find right here. List a set of filters on eBay. Even though you risk selling them for a low price if no one bids in a particular week, you can get a very high price if the bidding is active. I have sold sets of filters on eBay for as high as $150.00 and as low as $25.00. It's a risk you take.

If the bidding goes up to $125.00 to close your sale-you get to send a "second chance offer" to the other bidders. So you can sell another set to the guy who bid say $115.00, and another set to the guy who quit at $90.00. I have done this numerous times, selling multiple sets to several buyers off one auction.

CAUTION: Make sure you say in your eBay auction that filters take 2 – 3 weeks to ship because they are custom made, and a free test set of filters is sent first. You don't want negative feedback because someone expects immediate shipment upon winning the auction. Here is one of my auctions. Feel free to copy it.

Title of my eBay Listing

Microscope Filters Rheinberg Darkfield Lighting Optics

Notice all the keywords in the title, so anyone searching for any of these things sees my auction.

Pictures

Picture No. 1 should be a shot of a complete filter set, as close to what you are offering as possible. Then include more photos taken through the microscope, showing different Rheinberg effects, as described in the eBay listing. Include at least four or five additional pictures taken through the microscope using Rheinberg filters.

Ad Copy

Use text similar, but not exactly the same as the ad text I wrote below. Since you are writing your own ad copy, you will understand why I have made each statement. I use different fonts and different colors in the ad to keep people reading.

Text

These sets are available in any size diameter - the most common being 32mm for Zeiss standard. Measure your typical filter- such as a daylight blue, or neutral density, and let us know the diameter. Also, it would be helpful to know the maximum thickness your filter holder can handle.

Pictured, you see a complete thin plastic laminated filter set. You will not get this exact set. You will get a similar large assortment of our choice. You can mix and match center stop and annulus rings to give you more color combinations. You can overlay annulus rings to achieve different colors as well.

Are these high quality optical glass filters? NO. You would pay about $40.00 *each* for a solid optical glass filter.

This is an affordable collection for the serious amateur, university student, or even creative professional- in thin plastic assortment. You simply cannot purchase ready-made optical glass cut into rings in these colors.

Therefore- we highly recommend these- which are hand cut from various plastic laminates, filter materials, and adhesive films. We have made these filters for university professors, colleges, hobbyists, and scientists around the world.

What is in a typical set that you are bidding on? You get at least a dozen annulus rings. You get about a dozen center stop filters in different shades and colors. You also get several sizes of black center stop filters which give you

Darkfield for several objectives, depending upon your microscope setup. Usually, the bigger the center stop- the higher the objective power you can use.

[Picture Here]

To further explain about the rings: The overlays- or annulus filters- determine the subject color. These rings are in a wide variety of colors, and are stackable to increase density or create new colors.

Included are shades of green, shades of red, pink, orange, yellow, shades of blue, and combination-color rings. These combination rings also may be stacked to create additional colors, by offsetting the semi-circles at 90 degree angles to create quadrants of color.

[Picture Here]

Also included is a set of solid color filters. You will get a nice assortment of these with shades of green, shades of blue, pink, yellow, red, and a neutral density shade. Some people don't have these solid colors in their filter collection. If you don't have some of these colors- it's an affordable way to be creative.

Want to know more about Rheinberg and how to set it up? We include with your order instructions and tips on using filters.

For example: You must not stop down your condenser diaphragm. You need more light, obviously, and stopping down interferes with Rheinberg effect. Likewise, you must not stop down your base illuminator field diaphragm. No stopping down below the condenser- or you will lose the effect.

How the Rheinberg effect is created: The "center stop" determines the background color, and the overlay "annulus ring" determines the subject color. These are placed BELOW the condenser- in the filter tray. In this auction's photos, you see salt grains with the center stops of red and the annulus in blue. Cyclops microorganism also photographed the same way. A ring of daylight blue for natural subject color was used in those Rheinberg photos.

[Pictures Here]

A black center stop gives you a Darkfield effect. When used in combination with a light blue overlay ring, your subject has a natural color against a black background. The photo of Vorticella was done that way.

[Picture Here]

If you've never used Darkfield at all, then you are in for a treat because without Darkfield, you can hardly see many things. Most micro life is clear. Suddenly vorticella, paramecium, and all sorts of microorganisms pop out in brilliant whiteness against the dark background.

The Hydra photo was taken with green center stop and yellow overlay. The circular diatom photo was taken using a violet center stop and yellow annulus.

[Picture Here]

Technical Stuff: These center-stop and overlay rings work great using 4x and 10x objectives. Even using a 20x objective should give you very good results. But, you generally cannot use these with a 40x objective and higher because the precision needed in matching your objective and the center stop size is very critical and varies between microscopes and objectives manufacturers. If you want a Darkfield effect using the higher powered objectives, you generally have to use a special Darkfield condenser which is not part of this auction.

If most of your observations are in the range of 200 times magnification and below, then these are great filters for you.

The Daphnia photo was taken with green center stop, and daylight blue outer ring (annulus).

[Picture Here]

As you may know, bacteria are extremely hard to see, and usually require 1) fixing when dead, 2) special staining, and 3) very high power- say 400 magnification and up. Well, using Darkfield, you can see the tiny sparkling bacteria (many species) at 100X and 200X- LIVE! So, if you are new to all this, it will definitely take you to the next level.

Remember- be sure of your filter size, and let me know your requirements, as these are approx. 32mm in diameter.

NOTE: We cannot ship your custom filters until we send a sample set, at no charge, for you to test. Once you confirm the exact filter measurements, we ship your full color set.

NOTE- We can make custom colors and combinations as well- upon request, for example a ring in two colors specific colors- left and right. Finally, we will unconditionally guarantee these for you, for 90 days, as we want them to work in your particular set up. Therefore- if you need adjustments made to the size of the center stop, or the filter size, we will make those adjustments. Just let us know. If you have questions- just email us, and we'll help you work out details for your particular scope.

Do you want to try making your own Rheinberg Filters? There are lots of articles about How to Make Rheinberg Filters, and people use various materials and techniques.

My techniques are professional and I use precision equipment, and work with each person to ensure the center stop diameter matches a particular microscope. But if you don't have the patience to do it on your own- then this set is for you!

Thanks for looking at our auction!

Marketing through Groups

Join Yahoo groups about microscopy. On the main Yahoo website page, search the groups for "Microscope." Here is a link to the list, and some of the groups.

http://groups.yahoo.com/search?query=microscope

Some of the groups are as follows:

Microscope
Microscopes
CoolPixPhotoMicMac
Microscopehobby
Amateur_Microscopy
Wild_M20
Qx3
Micromounts
DarkFieldMicroscopy
Macro3D
AvianMicroscopy
Microscope_Restoration_Collection

...and many more.

Literally, there are thousands of members in these groups.

Pick some of the groups, and start participating. Always have a link to your website underneath, and your name in the signature area. You are not allowed to solicit in these groups; however you may say things like "check out the new photos on my website." That's what I have done to sell filters and many copies of this book. In a microscope group, it's okay to drive people to look at your website, and there, they can discover your Rheinberg filter page. I have been doing this for years.

That's just Yahoo.

Facebook has groups. Twitter has groups. Google has groups. Goodreads has groups. Join all the groups and be a contributor. This drives traffic to your website if you show the website address under your name.

One quick warning when you participate in groups, social media, and blogs: do not reveal your trade secrets if you plan to sell filters. You paid for this book, and the knowledge here is worth much more than the price of this book. You do not have to mention electrical tape, theatrical gels, or die punches, to contribute. You can talk about center stop sizes, photography, specimen collecting, annulus ring colors, placement under the condenser and so on, without explaining the techniques in this book.

Write Articles

There are many on-line magazines for microscopy as well, and they look for articles. If you love lichen, or protozoa, write an article about your favorite study topic. This will give you exposure and the opportunity to mention your web page. I've sold many filters simply by writing articles for the on-line microscope magazines.

Offer articles to blogs, and contribute to those blogs. Get on the internet and Google all the microscope topics and techniques. Any blogs out there should get a nice thoughtful contribution by you. Add your website at the end. That's not spamming, and it is actually helping people's blogs by giving them content: Namely your contribution. Be sincere. A Google search of forums will give you lots of places you can contribute and promote what you want to sell. Just type a forum topic name and the word "forum."

Join Clubs

By joining a local microscope group or several online clubs, you have the chance to network and meet new people who have a similar interest. That's a perfect venue for you to say that one of the things you do is sell Rheinberg filters, made to order. Many clubs have newsletters that are mailed out, emailed newsletters, or on-line newsletters. You can write an article for the newsletter, and it will be distributed to all of your fellow members. Nothing is stopping you from joining several clubs, including those for botany or butterflies, or other science related interests. Science clubs usually have members with microscopes.

Social Networks

Pinterest is the perfect place to post photos take through the microscope.

Facebook, Twitter, and other social media sites all have microscope topics and members who either love microscopy as a hobby or use a microscope professionally. I'll see you there! Once you are on Facebook, you can set up "pages" for your interests. Set up a microscope page on Facebook, and then you can build an audience. It's like having a free blog that give photos and video a huge audience on the internet. Open a separate Twitter account just for microscopy or Rheinberg Filters, then try to build followers. There are numerous ways to build a base of followers, the main one being to start following everyone with similar interests, and then asking them to follow you back. Every time you tweet, you will be in contact with hundreds of potential customers.

Once you have some of these marketing avenues established you will start to receive inquiries for your filters. Here are your next steps.

Responding to Inquiries

Here is a typical inquiry you might receive.

To: Mike Shaw
From: person@foreignaddress.countrycode
Subject: Rheinberg filters

Dear Sir-
I would like to purchase a deluxe set of your filters for my XYZ brand microscope.
Can you please tell me the price and the cost to ship to Mycountry?

Thanks!
Ms. Filterfan

Notice that the information she provides is very scant. That's normal. Most people who want to buy filters don't give you a lot of info, and if you are lucky they may tell you the type of scope they use. They almost always ask for the price.

If they say Zeiss Standard, then you know you probably will be making 32mm filters. But you always have to confirm filter size. There are a few extremely important questions to ask.

Below is my typical reply tailored to the specifics of the potential customer. If the inquiry is from overseas, I always quote the foreign postage, and also add $20.00 to the sales price. This is because the free test filters cost more in postage, and sometimes they must be tweaked, and mailed twice. Since I never charge for test filters, I'm covering all potential costs up front.

> From: Mike Shaw
> To: person@foreignaddress.countrycode
> Subject: Rheinberg filters
>
> Ms. Filterfan-
>
> Thank you for your inquiry. I am assuming you saw my website www.rheinberg-filters.com Can you please confirm how you heard about my filters? I'm always curious to know.
>
> The procedure is that I send you a free test set of filters for you to keep. We use this to precisely determine the right size of filters for your scope.
>
> I will need to know the type of scope you have, and whether the condenser has a swing out ring for filters. This ring would be attached to the bottom of the condenser. You can see how it looks on my website.
>
> If you can email me a digital photo of the bottom part of your scope, showing where the filters go, that would make it even easier. Or, you can just describe how the filters are placed.

I also must have the exact size of your standard filter. Measure the diameter of the filter that came with your scope, in mm, as precisely as you can.

The price for a full set of Rheinberg filters is USD $100.00, plus USD $20.00 for international airmail and handling. You can pay via PayPal after your full set of filters is complete and ready to ship.

The full set includes over 15 center stop filters, approximately a dozen annulus rings, a set of solid color filters that match the annulus rings. You could use these solid filters with certain specimens, and add lighting from the top.

Please let me know if you have any questions, and thanks again.

Rgds,
Mike Shaw

Feel free to use my wording above, if you like. The key elements in your response should include the following. These are the basics of any business deal, incorporated into a friendly email interaction.

How did you hear about my product?
You will receive something free with no obligation.
What are the specs of your microscope?
The price and shipping cost.

Payment method and when due.
What you, the customer, receives for that price.

We will look at each point individually. Your goal here is to get a commitment to buy. You have not sold the filter set yet. When the person responds with the information you have asked for, they have almost committed.

What seals the deal is when they give you their mailing address. You do not ask for this up front, because this might scare them off. If your initial response said, "Here's the price, give me your mailing address," they might back off. It's frightening to give your address to a stranger on the internet. So you must establish some rapport, show you are legitimate by taking an interest in their microscope and providing facts about how you operate, and make clear that you don't expect payment too far in advance, but only after the filters are competed and ready to be shipped.

When the customer responds with everything you asked for, then you send a short email saying, "Great. I will start on your sample set this week. Please provide your mailing address and I'll send you them as soon as they are finished."

It's a small, but important detail. Look at it from the customer's point of view. They will ultimately be sending you around $100.00 before they have received a product from you. So, you have to build trust, and you do this by back and forth email, and by promptly sending out the free test set. Now let's analyze the email exchange a bit more.

How did you hear about my filters? You want to know what is driving traffic to you. Is it word of mouth through some club, or is your website coming up in a Google search? That's something you want to know. Whatever is working, you want to do more of it.

You will receive a free set of filters (no obligation). The free sample set is important for two reasons: technical and financial. On the technical side, you do not want to make a full color set of filters in the wrong size. On the financial side, you need to overcome any customer resistance to sending you payment prior to shipping. You do that by making sure the customer knows that these samples are his or hers to keep. Occasionally, people will ask if they should send the test filters back to you.

Payment method and when due. First you have made sure that the customer feels that they are getting something of value up front, at no charge: the sample set. This establishes your credibility. Next, you mention that PayPal is the preferred method of payment. It protects you the seller, and it also protects the buyer who can dispute you if he is not satisfied. Avoid receiving checks, which you have to deposit in your account, and thus you are giving your account number to a stranger who sees it on his canceled check. Never, never accept cash payments, as someone can claim they sent it and it must've been stolen. Insist on PayPal, if you can. Never allow the customer to pay after you ship, because that means either he doesn't trust you or he doesn't have the money. Either case does not bode well for business.

The buyer's specifications. Nowadays, I almost always get an email with pictures of the customer's condenser and microscope base. Even so, you must get the customer to tell you the exact size of their standard filter. If they can't tell you that, then chances are they won't be able to figure out how to use your test set, even if the test set fits in their scope.

If the customer does not have a standard filter to measure, then they must use a caliper or ruler to measure the inside of their filter holder. I have had customers do this too. You must have as close an approximation of the test filter size as possible. Otherwise you will be making multiple sets of test filters until you get it right. I will deal more with this outer diameter issue when we get to the following section on sending out the test set.

Where is the filter placed? That is a super important question. As discussed in both the requirements and testing sections of this book, if the filters are placed on top of a hot base illuminator, you will have to decline the order. If your filters can be placed directly under the condenser in a slider or in a swing out filter ring, that is what you are hoping to find out.

Price and shipping cost. You save this for almost the last item. You should never lead with price, not in any sales situation. You must always show value, and how your product fits a need. Then, you can reveal the price, at which point it is understood in its true perspective. What should you charge?

Set your price anywhere you like, but it has been my experience that it is not worth the time and effort for me if I don't get at least $75.00 to $100.00 per set. Maybe your time is worth more, maybe less. I've found that the average microscope hobbyist, your main customer, can't afford more than a hundred bucks. Even though your time may be worth more per hour, your filters are only worth what someone will pay for them. Remember--in most cases there is a spouse or partner that has to be sold on the idea. Anything over $100.00 can be a deal breaker for a wife who's been waiting for a new garbage disposal.

And don't think you can charge a lot more if some university professor or scientist wants filters. Finances are tight in the science world, often dependent upon hard won grants, scholarships, internships, and such. A teacher or professor often pays out of his own pocket for the tools he or she needs. So don't let a fancy title or position lead you to overcharge. I sold filters to the Mayo Clinic at my normal price. I'm happy to get the bragging rights. Try to be consistent with your pricing, except in one area.

Give some people a break. If you sense that someone really cannot afford to spend a lot, you can drop your price out of sheer kind-heartedness. If someone says they can't afford a full set, and they ask how much a set of six filters is, you can be generous at that point. It takes you almost the same amount of time to make twenty filters as it takes to make ten. So you may as well lower your price if you want to sell a set, and count it as an act of generosity. Make and send the full set.

On the other hand, if someone even hints that your filters are overpriced, or argues about the price, stand firm. They don't realize that your filters are in fact underpriced for the labor involved. It takes a big part of your day to do one set of filters.

Postage is a wash. Don't make money on shipping. Filters weigh very little, and take the minimum Priority Mail rate, and the minimum International Priority Airmail rate. I don't even ship them Parcel Post anymore because I want to know if they arrive, and I want to know that immediately. Priority Mail always gives you a Delivery Confirmation.

What the customer receives for the price. You've revealed the price, and now you must quickly remind the customer what he or she is buying. That is the single thought in every buyer's mind the moment they know the price. "What am I getting for that price?" So you answer that thought.

Tell the customer he or she is getting a set of approximately "X" number of center stop filters, "X" number of Annulus filters, and an additional set of solid colored filters. This "X" quantity depends upon how many colors you have stocked up on, and what size of center stop the customer needs. Allow me to point out again that if the center stop size exceeds 3/4 inch, then you are limited in colors. There are only a few colors in rolls of tape that exceed 3/4 inch. You are stuck with scraps of signage material, and not all of that is translucent enough for good center stop material. Unless you have a dozen rolls of Con-Tact® paper in different colors, you will need to tell the customer as soon as possible that color choice is limited. Explain that due to the size of his center stop requirements he will only receive "X" number of center stop colors.

The reason I promise an additional set of colored filters is that you can punch these out at the same time you are punching out the disks for the annulus rings. We know that the cost is negligible. So by adding a little extra effort, you provide a whole additional set of options to customers. They can use a solid colored filter, and illuminate a specimen from the top with an LED or fiber optic cable. In sales this is called "Value Added." That means the customer gets something additional, without you incurring extra cost.

Now that we've analyzed the basic content of messaging, let's talk about service.

Response Time and Reputation

Have your canned reply ready. Save it in a Word document or in an email to yourself. Paste it into your reply, and then tweak it. If you are ever going to sell filters, or anything else, you absolutely must provide the best customer service and response time on the planet. I run over a dozen Amazon stores, sell filters, e-books, and sell items on eBay. If you take a look at my ratings you will see that my feedback is always at 100% positive. This comes from quick responses. I cannot stress this enough.

Along with customer service, you must be prepared to walk away from a deal that could make you look bad, or that you can't fulfill. That, too, is a response, and it must be a quick decision. These challenges do not come along every day, but once in a while, you will be up against the wall. Let it go, and if you have to, lose money and cut your losses.

A gentleman in Austria bid the highest price on eBay and won a set of my filters. It clearly says in my eBay listing to allow 2 weeks to determine correct sizing, mailing of the test set, and so on. Well, this customer was super impatient, emailing me every day, and telling me he had a big microscope meeting coming up, and saying he wanted his filters immediately. He just would not accept what I was explaining about how important it is to know the correct size of filter center stop.

Looking back, it was probably more important for him to show off his new state of the art Rheinberg filters to his microscope friends than it was to have an actual set that worked in his microscope. He was the customer, and he wanted his filters, for whatever reason.

I was not going to risk bad feedback on eBay from an impatient buyer who won my auction. So I kicked into gear, and with the information I had from him, and I cranked out a full set of Rheinberg filters, and shipped it out right away. You know what's involved, and you can see that it takes a few hours for each step of the process. But I quickly made his set, and shipped it, knowing full well that if it wasn't perfect I would have to do another set at no charge.

It's more important to satisfy the customer, and keep your ratings high, than worry about losing some of your time, and a bit of money. If you simply cannot get a test set of filters to work perfectly, don't make the color set. Issue a refund if they paid on eBay. Apologize, and recommend that the customer buy either a different microscope, or a Darkfield condenser made specifically for his scope. If the only possible filter placement is on top of a hot base illuminator, apologize, and then move on. Some microscopes simply will not achieve a Darkfield or Rheinberg effect with a center stop filter. There's no way to fix this. Don't send something which "may work," unless it is a free sample set.

When I first started, I used to state in my auctions and on my website that my filters are guaranteed for 1 year. That's plenty of time for someone to see if they melt or if they work. If, however, two years have gone by and someone wants a refund, I would still give the refund. You can't afford bad press from even one person who has access to all the same websites, groups, and blogs that you have access to. Uphold your reputation by being more than fair, and always leave every customer happy to have dealt with you. I have never had a set of filters returned.

Keep Excellent Records

I print every email chain when I start on a filter set. I'm not saying you have to do this, but you must keep good records. Sometimes a customer will come back to you for another set of filters for a different scope, or another set for a new objective lens they just bought. Keeping good records allows you to build repeat business by showing a customer that you remember him or her.

I've had a number of people win my filter auctions on eBay, yet never respond to me with the test results. This is a down side issue when listing your filters on eBay. One buyer did not respond after receiving the test filters—until he contacted me two years later. He was ready for his filters, and asked if I remembered him. I looked through my files, and sure enough, there was an auction sold, not shipped. We renewed our discussion of the test set, and he said he would do the tests now. I promised to immediately make his full color Rheinberg set, once he gave me the results on the test set. I did not hear from him for another two years. That's right. Total time elapsed—four years. Again, he asked if I remembered him and he promised to do the tests. That was a year ago as of this writing. I am still waiting to finalize this eBay sale. I will make his filters whenever he is ready, I suppose.

A deal is a deal.

Fulfilling the Order

You've sent and received emails, you know the outer diameter of the customer's filter, and you have a mailing address.

Now it is time to make up a test set. This consists of a few outer diameter test disks which vary in size by about 1mm. You want to be sure of the outer diameter too. Then, you include a complete test set of center stop disks, about 6 to 8 in various sizes. The test filters should be in a small Ziploc bag. Do not send them loose in the envelope. You can use the small Ziploc snack bags which you'll find in the grocery store, or use the very small 1 inch x 2 inch Ziploc bags that are available commercially. Get these at ULINE®:

www.uline.com

With each test set, send an instruction sheet, using the instructions I have shown in Chapter 5, The Filter Making Process, where I have on page 108 the sub-heading Testing Your Filters.

Once the customer receives the test set, it will take a week, sometimes a month, and sometimes years before they run the tests. You will find people very eager to get the free test set, and then suddenly they lose interest. People get bogged down in their various schedules, and somehow, the work of testing filters gets put off. I can tell you quite seriously that about 20% of the people to whom I send test sets never reply after that. So accept it, and don't worry.

Because I keep records of the correspondence, after two months I always send a reminder, asking in a very nice way if they have had a chance to run the tests yet. I then add that there is no rush or pressure, and whenever they get to it, just let me know, because I will save all of their info. As I've shown you above, it might take a year to hear back. If I don't hear from someone for another six months, I'll send them a very similar reminder. I'm doing this just in case they lost my email address. Maybe they bought a new computer, or maybe they moved. You never know. After a year, I give up, and move on.

During this time, you will have probably had other customers to keep you busy. You should have sold enough sets so that you will not lose any sleep over some test sets in limbo.

After customers test your filters, they will always come back with a few more questions. Although they may need guidance from you to help them decide on which center stop size seems best, always let them make the final decision. A customer will have conducted the necessary tests with your sample set, and it only makes sense that he or she settle upon a best choice in an email. If you don't get this agreement, and your decision doesn't work, then you might find yourself making another full set of color Rheinberg filters for free. Avoid that by letting customers be sure of their own decisions as to which center stop size works best for them.

Once you know precisely the outer diameter of the customer's filters, and the exact diameter of the center stops that need to be made, you can begin making the full color set

First, punch out your main colors for the annulus rings and the solid filters. You simply punch out two disks of every color you have, in the correct outside diameter for that particular customer. Then, punch out the inside holes for the annulus rings. The inside holes will be the same size as the center stop size. If you make a mistake and the hole is off center, then punch out another disk and do it again. You cannot send a customer an annulus ring that has a hole that is even slightly off-center. It has to be perfect. You will wind up with a set of solid color filters and a set of matching annulus rings.

Next, punch out a set of center stops, in the exact diameter agreed upon with the customer. Once you have the center stops ready, then punch out enough clear disks to match the number of center stops, as well as a few extra ones, because you might have to do some over.

Using a tweezers, carefully lay the center stops in place on the clear disks. If you make a mistake, you can sometimes peel off the center stop, and reposition it correctly. If you can't peel of the center stop easily- you have to do this one over. Placement and centering of the center stop is a skill you will develop as you make more and more filters. If you do make a mistake, then just punch out a new center stop and place it on a fresh clear disk. Do not re-use the old disk, as some adhesive residue from the incorrect positioning might show up under the microscope.

The customer's Rheinberg set is finished.

Sometimes a customer will want two sets of filters—for two separate objective sizes. Since you won't have to make an additional set of solid color filters (because the outer diameter for the additional set is the same as the first set), you can offer a discount. You are already making a set of filters for this customer, and there isn't that much more work to crank out double the number of clear disks and an additional set of disks for the annulus rings. Certainly, the additional cost in materials is negligible. I usually offer that second set at 1/2 price. The customer is getting a great deal and you are only doing a bit more work. Your shipping will not increase, and the filter case cost is the same.

Because you are charging a premium price for a set of custom made Rheinberg filters, it is important to ship them in a nice case. This is a value the customer deserves, and also shows a certain amount of pride you take in your work. It is the professional way to go. In Chapter 8 Cases, I talk about various cases. My favorite case is the multicolored, cylindrical tower, with separate compartments for each color of the rainbow. Chapter 8 tells you where to get these colorful cases.

I stack the colored filters, annulus rings, and center stop filters in their respectively colored chambers. Using Kimwipes®, I tear about 1/3 of a sheet, and pack the crumpled paper on top of the filters so they do not shake during transport. See the Tools chapter for info on Kimwipes®. You will also need to line the bottom of each compartment with a disk of craft foam. See the Materials chapter for info about craft foam. You want to protect the filters from scratching, and the bottoms of these compartments have a small center bump that could warp the shape of your filters.

Include the Rheinberg Filters instruction sheet that you will find below. Include the extra lid that comes with the colored filter container, in the event that the customer wants to split the tower into two stacks. The extra lid allows that. Do not include the blister packaging that the tower comes in, shown here. Most of the time, your buyer will never have seen a case like this, and will be very impressed. I never tell the customer about this rainbow case in advance, not on my website, not in eBay ads, nor in any email. I want it to be a surprise. There is a saying in sales, "Under-promise, and over-deliver." When the customer sees his set of filters arrive in a beautiful case, he or she will be blown away, I promise you. So don't downplay it by including the cheap-looking store blister pack.

Instruction Sheet Sent with Each Filter Set

Rheinberg Filters Instructions:

Note: These filters are designed to work with objectives of 20x and lower. They work best with a 10x objective or with a 4x objective. Higher powered objectives (such as

40x and up) require high precision measurements or a Darkfield condenser.

The thin clear filters with a solid color "center stop" determine the background color. The center stop filters with black background are for Darkfield. Several Darkfield sizes have been previously sent to you as test filters for your experimentation to see which objective works best with which size center stop. Those are your Darkfield center stop filters.

The colored "annulus rings" give the subject its color. (The annulus ring is a ring of color that will surround the center stop.) Use a light blue ring for a natural subject color, like daylight.

First, place a slide on your stage, with no filters in place, and focus on the specimen. Select a specimen like salt crystals, diatoms, or any subject that normally would be washed out and almost transparent. Animal hair, strands from a cotton swab, and crystals are also good subjects. Focus on the specimen. Condenser should be racked all the way up, almost at the top, if not completely at the top.

Next, select a filter that has the center stop color of your choice: black, blue, red, or green, for example. The black center stop is opaque, and will give you a Darkfield effect. Place your choice of center stop filter in your filter tray.

Then, select a contrasting colored ring (annulus ring), and overlay this on top of the center stop filter.

This two-filter combination should be in a filter holder directly under the condenser.

Important: Your condenser iris or diaphragm must be wide open. Do not use a swing in lens under your condenser either. You may have to raise or lower your condenser very slightly to get the best effect. If you have a field diaphragm (in the base illuminator), this iris must also be wide open.

Achromat condensers and objectives work best. If you use phase-contrast or plan objectives, or a 1.4 NA condenser, you may not get as good a Rheinberg effect as you will with a less expensive condenser and an ordinary objective. Neofluor or Plan Apo objectives don't give the Rheinberg effect as well as cheaper achromats.

These filters are plastic. Do not place filters on top of a hot illuminator. They will warp. Turn off your illuminator when you are not looking through the microscope, to keep the filters cool (unless you have LED illumination).

Mix and match the various colors to create unique effects. Although some colors may look the same- they may be several shades different. These vary slightly under the microscope in hue and density. You can also combine two of the same color family to give a darker hue.

To be sure that these are working perfectly—you need to see a nice evenly colored background across the entire field. And your subject should light up like a bright field of stars evenly across a winter midnight sky.

TROUBLESHOOTING: If you are having problems- try using only the center stop colors, without any annulus rings. If the center part of the field is much darker than the edges of the field, and your subject (like diatoms) is not very bright in the center- then you have one of several possible problems:

You are using the wrong size objective. Use 4x, 10x, up to maximum 20x.

You are using a Neofluor or Plan Apo, or another high end type of objective. Use an ordinary Achromat instead.

Your condenser is too low or too high. Make slight adjustments.

You stopped down the diaphragm in the condenser (iris), or in the illuminator (field diaphragm). Leave it wide open.

You have the swing out lens under the condenser (the space should be clear between the filter and the illuminator).

The center stop of the Rheinberg filter is the wrong size for your particular objective, or microscope, or condenser.

Go back to your test set, and worksheet. Start over by testing each size Darkfield filter with your objective.

Remember - You get a 100% satisfaction guarantee for 90 days, and a guarantee on the materials for one year- free replacement!

If you have any questions at all, or any concerns, please contact me: yourname@email-domain.com.

Shipping the Filter Set

Send an email to the customer saying the filters are ready to ship, and invoice them through PayPal. As soon as you receive payment, ship via Priority Mail.

The case and instructions easily fits in a small sized Priority mail box for a fixed rate of around $5.00 for USA. Make sure you ask for Delivery Confirmation when you mail it, and you do want to make sure your labors are not in vain by confirming the customer receives the filters.

As soon as you ship the filters, you need to advise the customer via email that you've just sent out the filters by Priority Mail, and provide the Delivery Confirmation number. If I don't get a "Thank you, the filters arrived, and they are fantastic," email in a few days, I track the package and make sure it was signed for. Then I wait a couple more days and ask the buyer if he got his filters. I want to be sure, and I want him to acknowledge that fact.

If you get a really nice thank you, make sure you ask for permission to use that as a testimonial on your website. Then edit the name and email address, and post it on your webpage. In all cases, reply to a thank you with "Please let me know if you have any questions." In most cases, if you have done everything correctly, you will never hear from this customer again. In some cases, they will come back to you with another filter order for a different objective lens size, or for a different scope. Don't underestimate word of mouth, as you may get new orders from referrals by happy customers.

That's it. You're in business! Go make some money and have fun with it.

CHAPTER 10

How to Make a Microscope Camera Adapter

The adapter does not have to be fancy or expensive. It just has to be more stable than your hand held camera. The best place to buy your parts is at the hardware store. The big chain stores or even your locally owned store should have everything you need.

This chapter is intended to help you make an adapter for a point and shoot digital camera. A heavy DSLR (Digital Single Lens Reflex) really should be on a tripod, and then centered over the microscope eyepiece. Most microscopes aren't really designed to support heavy equipment sitting on top. Besides, DSLR's have a threaded or bayonet front, on which you can attach a commercially made adapter. Since you would have to purchase an adapter specifically for your DSLR camera model or lens, and possibly a series of step rings in between, that's not what this chapter is about.

The good news is that the adapter about to be described may also work for your DSLR, if you really want to forego buying a commercially made adapter and the step rings.

Here is a picture of a commercially made microscope adapter. The bottom part has the thumbscrew which tightens on the barrel of the eyepiece tube. The top is threaded for adapter rings which will to fit the camera.

A point and shoot digital camera typically does not have anything on the front, threaded or bayonet, onto which you can attach a stable microscope adapter. This is because point and shoot cameras are meant to be compact, and the lens retracts into the camera body leaving a flat surface. So the idea in this chapter is to build an inexpensive adapter. You will simply drop your camera into it, and have the stability you need.

In fact, with an inexpensive homemade adapter, you will be able to shoot great microscope pictures with not only an inexpensive point and shoot camera, but possibly with a lightweight DSLR, a webcam, and even your smart phone, or tablet.

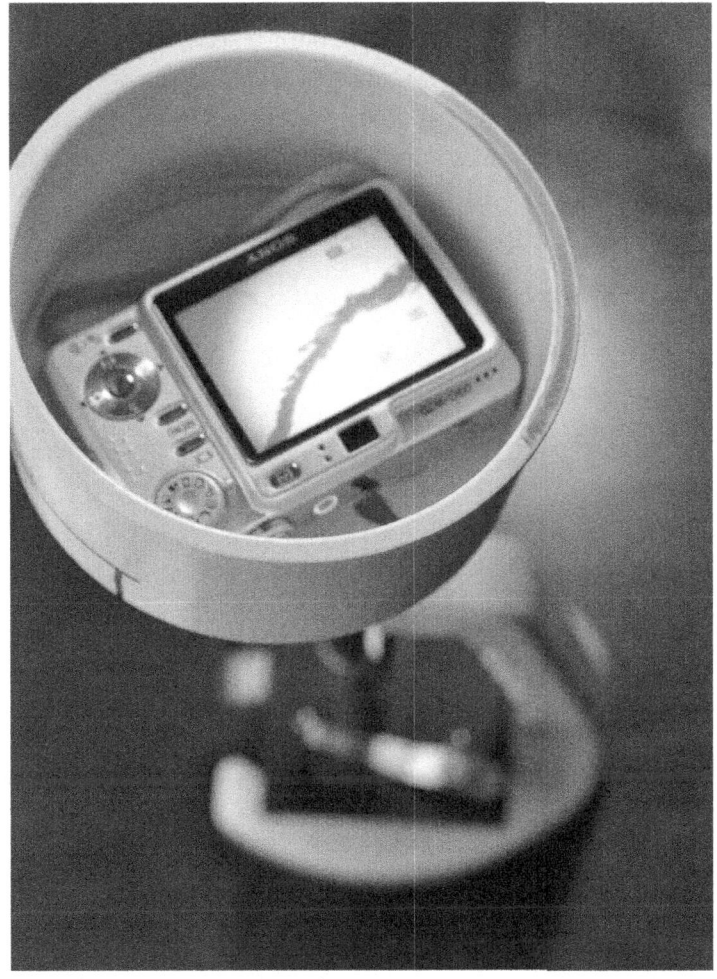

My camera for the above set up is a Sony Cyber-shot 7.2 mega pixel point and shoot. The image is of the slide with the piece of moss which was used in the previous pictures.

Before you begin buying the components for your adapter, you need to make sure your camera can take pictures through your microscope. Start by setting up a nice colorful specimen on a slide, perhaps a strand of moss, or a strew of colored sugar crystals. Try to cover the whole field of view from edge to edge with your specimen. Focus the microscope, and make sure you have a perfectly sharp image using your eye to the eyepiece. Below, I'm using a Nikon Coolpix S8200 for the test just described.

Your Microscope Hobby

Turn on your camera, put the setting on macro, and either set it up on a tripod or hand-hold it over the eyepiece. See what kind of image you get, and then if need be, zoom in a bit. Try to zoom in until the circle of light fills the entire camera LCD screen with your specimen image. If you can only get a small circle as in the above picture, filled with a small image of your specimen, then that is all your camera can do. Consider buying another camera if you want a bigger image. Sometimes a cheaper camera works better (the Sony). Directly above is a top of the line Nikon which is not as good in terms of image size when compared to the less expensive Sony Cybershot.

A more expensive camera will not necessarily give you the best image in the view finder. It is more a question of optics than a question of price.

When testing your camera, hold the camera as close to the eyepiece as you can, without actually touching it. It is best if you use a tripod to to do this (as I did here with the Nikon S8200). I slowly lowered the camera into place, keeping the image in the center of the screen.

Now take a measurement of this ideal distance between the front of your camera lens and the top of the microscope eyepiece. It should be only two or three millimeters. If it is too small to measure, then that's okay. Just make a mental note of it.

Your Microscope Hobby

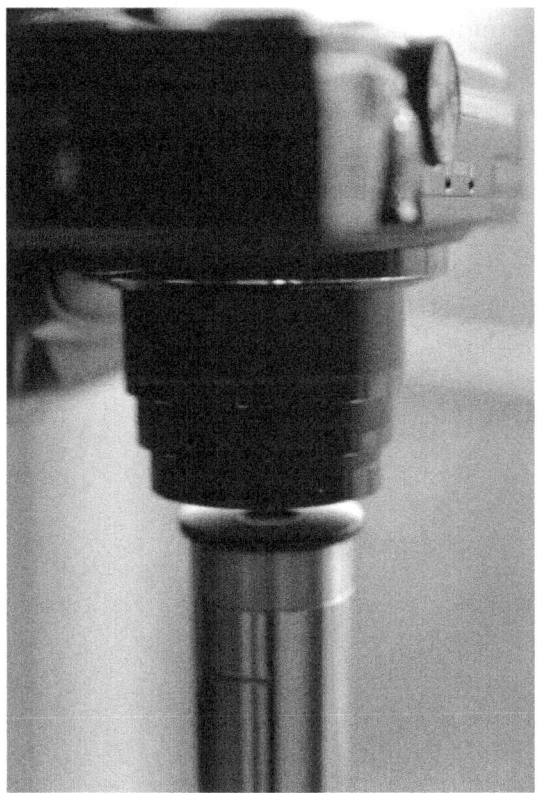

Microscope Eyepiece Tube Measurement

Microscopes have standard eyepiece and tube diameters on the outside. That's all you need to be concerned with. The outside diameter is usually about 25mm or about 0.98 inches. Therefore, a PVC pipe with a 1 inch inside diameter will hold an eyepiece tube, but not the edge of the eyepiece. This is because the eyepiece has a slight overhang or flange which prevents it from falling into the tube.

Therefore, do not remove the eyepiece from its tube. Measure the outer diameter of the eyepiece with its flange. Use a caliper, or use a precision ruler. For example, you can place a ruler across the top of the eyepiece tube. Mine measures 28mm. You can also trace a circle on a piece of paper. You can even look around the house for a medicine bottle of the correct diameter. My medicine bottle was a perfect fit. You will need this measurement to bring to the hardware store.

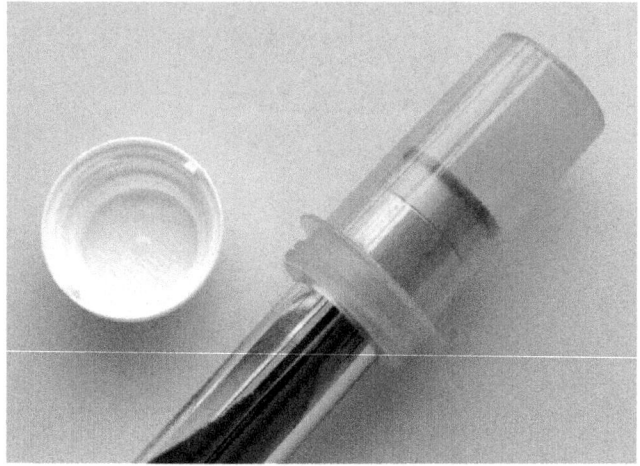

Camera Adapter Tube Measurement

Now turn on your camera and let the lens pop out as in the photo below. Measure the outer diameter of the camera lens at its widest point. You must also measure the length that the lens extends from the camera body, as in the below picture. Important—put your camera setting on *macro* because that is what you will use when taking pictures. The macro setting may affect the length that the lens extends from the camera body.

If you are using a DSLR, measure the widest part of the lens, where the zoom or focus ring is. Be sure to use the lens you will use for shooting macro shots. Also when measuring the distance from the camera body to the front of the lens, be sure to include the UV filter, which you should have on the lens. Always protect your DSLR camera lens with a filter.

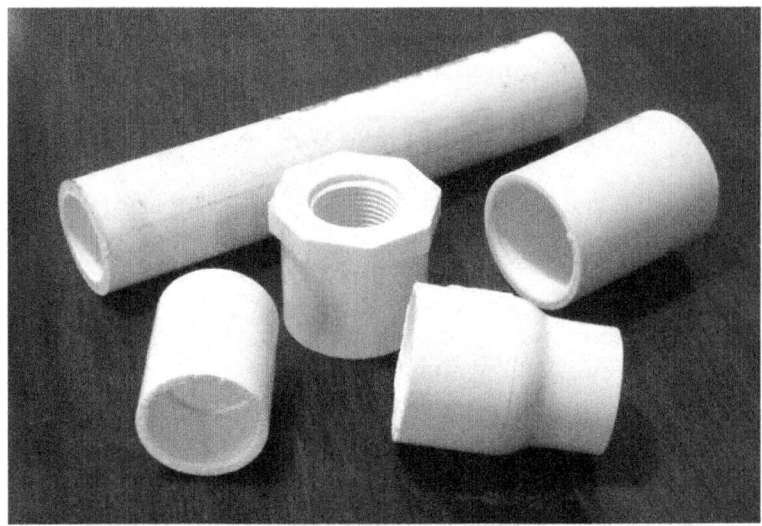

Your entire adapter project should be made of plastic, so it will not scratch the microscope or your camera. PVC plumbing parts are the best solution for this. For looks, yes, there is nothing like a polished piece of brass or copper (well, gold is nicer), but anything made of metal will scratch your microscope. Therefore, plan on a crude makeshift look, and spending around ten bucks.

Selecting your Components

You will need a big piece of PVC for a camera cradle. Look for a large *end cap*. Mine is a 4 inch inside diameter end cap which will eventually fit over a large PVC pipe.

You then need an eyepiece tube with inside diameter slightly bigger than your microscope eyepiece tube. I found a piece with an inside diameter of 28.2mm. It is designed to go on the outside of a 1 inch PVC pipe.

Finally, you need a camera adapter tube for the camera connection to the eyepiece tube. You will have to get something that fit over your camera lens, and maybe another tube as well to help adapt this down to the narrow eyepiece tube. I was able to forego that extra adapter step-down tube because one single fitting worked fine for me.

Additional parts:

- 3 plastic or nylon thumbscrew knobs
- PVC cement.
- Plastic model putty. Wood putty is okay, but get the kind that will adhere to plastic.

Tools You Will Need

- Eye protection (goggles or safety glasses)
- Screw tap that matches the plastic thumbscrews
- A drill (electric or hand drill)
- Jig saw or band saw (a hand held jig saw is fine)
- Vice
- File
- Sandpaper
- Vacuum cleaner!

Here are some suggested sizes and guidelines for the components and tools:

For your camera cradle you need the biggest PVC platform you can find. Your camera will rest on top of it. I used an end cap with a four inch inside diameter. Depending upon the type of drill and saw you have, you can really buy anything you feel you can deal with. Remember that you will need to bore a large hole into the end cap, and you will need to cut the sides off of it. So have fun picking out your camera cradle. Below is the finished adapter, showing the end cap I used for a cradle. Black lines are where I can use a jig saw to trim off the sides of the cradle.

As said earlier, I was fortunate enough to find a camera adapter tube that was the correct size on both ends, so I could use the one component and no additional tubes. I attached one end to the end cap, and I was basically done. The end attached to the end cap held my camera lens perfectly.

The other end (about 1 inch inside diameter) was able to fit over my microscope. My camera for this set up was a Sony Cyber-shot 7.2 mega pixel point and shoot. Try out your parts in the store. Look for adapters and bushings.

Your Microscope Hobby

You may not be so fortunate, and you might have to buy separate tubes and interconnect them. If that is the case, you will have to do quite a bit of testing one tube with another, as in the below photo.

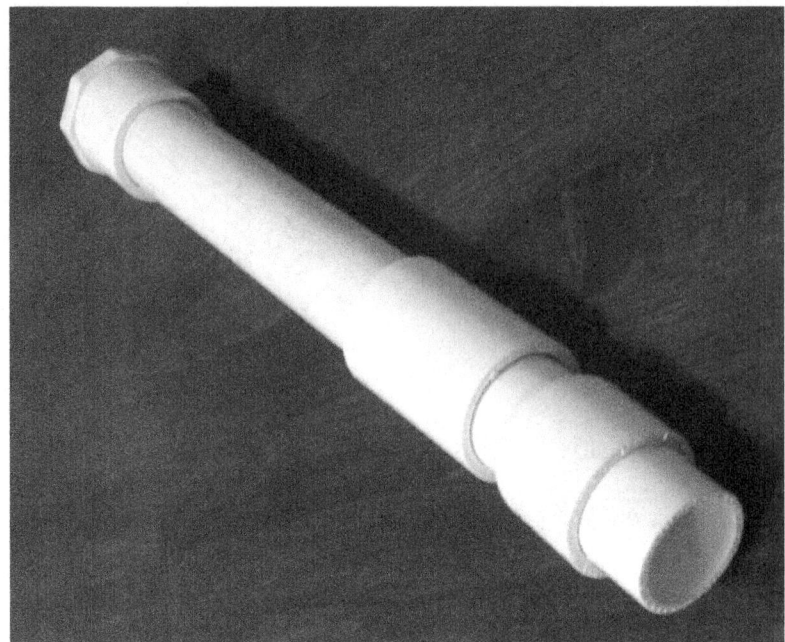

The goal is to have your microscope eyepiece slide smoothly inside one bottom tube, and (maybe by nesting that into another tube or series) have the other end receive your camera lens. Your camera lens will drop into the top part, and slide smoothly into it, coming in close contact with the microscope eyepiece. Cut all tubes down as short as possible so they work and are stable as well.

For your microscope eyepiece tube, the inner diameter of this tube must allow a free sliding, not too snug, fit over the eyepiece tube *and the eyepiece flange*. See below picture. Buy a length of at least three inches in case you need to trim some.

Your Microscope Hobby

Now, for your camera adapter tube, as suggested, you may need to find possibly more than one piece to connect between camera lens and the eyepiece tube. So select your microscope eyepiece tube first, and then try to fit it into a pipe fitting that has an opening the size of your camera lens. There are many such PVC pipe adapter fittings already available, and no modifications may be necessary.

Bring your camera to the hardware store, and try it out with some plumbing parts. Not to scare you, but you may have to make several trips to the store, returning the items that do not work. Save your receipts. Once you find the winning combination, you are done.

Find plastic or nylon thumb screws in the thin sliding drawer trays in the hardware aisle of the chain stores.

In smaller hardware stores, just ask for them. The size I use is 8-32. Here is a picture of the nylon thumb screw.

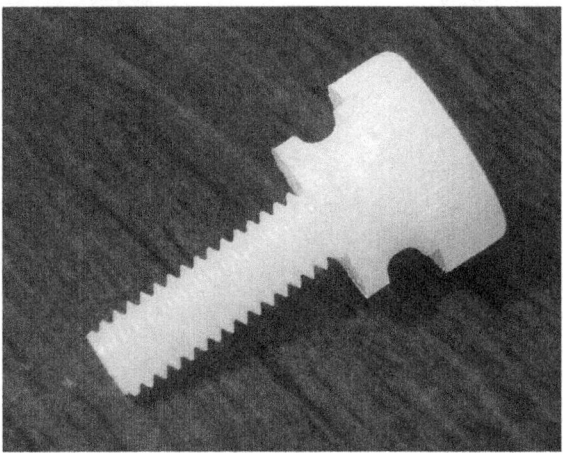

If you don't already have a screw tap, buy one to match the thumb screws. It's a slight investment; however you will be able to use this tool in the future. If you've never used a screw tap before, don't be intimidated. It was new to me as well. It's great fun to learn how to use a new tool, and it is easier than you think. Mine is size 8-32. The screw tap comes in a set with a matching drill bit for making the hole.

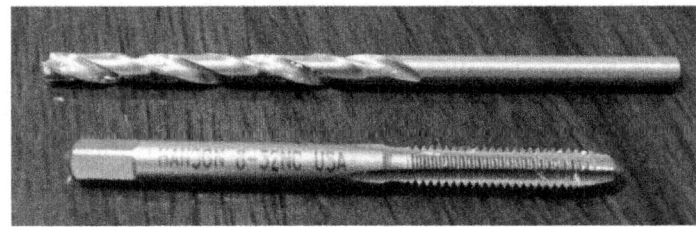

Buy the smallest can of PVC cement you can find, and I recommend you buy the kind that does not require two cans (A & B) to work. PVC cement is messy, smelly, highly flammable, and toxic. And when you are done with it, you must bring it to your local town dump that accepts paint related materials like this. Do not throw it in the trash, because it is very bad for the environment, and is a safety hazard. After you buy it, don't leave it in a hot car. Drive straight home with it and put it in your garage, and don't put it anywhere near a hot water heater.

Plastic model putty will adhere very well to PVC plastic, and this comes in small tubes which contain just the right amount for your project. You can find this in a big crafts store, and most certainly in any hobby shop for making model planes and cars. You can also use wood putty, as long as it is plastic based and will adhere well to PVC plastic. You can read the back label, or test some on a piece of PVC and see if it sticks. You will be using the putty to seal any cracks between the various components. Of course, when you are done you can sand and paint your masterpiece, and your adapter might look really great.

Building Your Adapter

Drill a large hole, off center, in the bottom of your big camera cradle. You have to decide where to drill the hole, based upon the position of the lens in your camera, and how it will sit in or on the cradle. The idea is to be able to rest the camera in the cradle, and have the lens hanging down through it.

The below picture shows where I drilled the end cap, inserted, then cemented the microscope adapter tube.

The size of this hole will be exact size of the outer diameter of your PVC camera adapter tube. The adapter tube must fit very snugly through the hole, so the PVC cement will work properly. You may have to drill the hole slightly smaller and then file and sand it outwards to the correct tight measurement. When you drop your camera into the cradle it should line up with the hole. You may also cut away the sides of the cradle with your jig saw or band saw to make your camera fit properly.

As explained above, my project was almost finished after cementing in just the one adapter tube. The top inside the cradle allowed a perfect fit for my Sony Cyber-shot camera to drop in, and the underside of the tube fit the microscope eyepiece. See below picture.

Your Microscope Hobby

You will have to work out the various tube combinations for your particular camera. You may have to buy another camera just for your microscope photography. Run tests by taping the parts together, and then when you have it right, apply your PVC cement per the instructions on the can.

This is a trial and error process; however with patience and ingenuity, it will work. Use the putty to fill in any crack or gaps. Smooth it out, and when the putty is dry, sand it. Let it dry in a well-ventilated place over night.

The last step is to drill three small holes, using the drill bit that goes with your particular screw tap. These three holes will be evenly spaced around the bottom end of the microscope adapter tube. Drill these holes about 3/4 inch from the bottom of the tube. Below picture shows drilling the hole in the tube.

Now tap out the holes using the screw tap. You'll wind up with threaded holes like the one shown below.

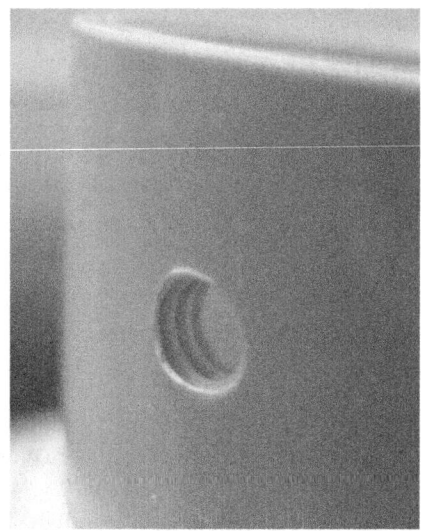

Next, insert the plastic thumbscrews, screwing them in just enough to hold them in place, but not past the inside surface of the camera adapter tube.

Your Microscope Hobby

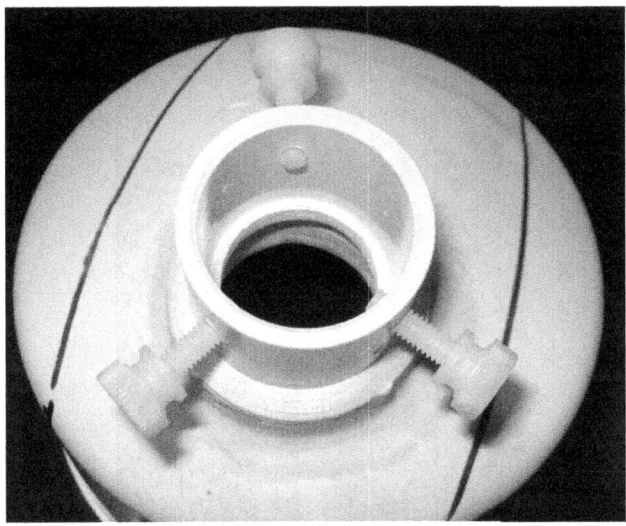

The black lines show where I could cut the end cap sides to have more access to the camera controls if I wish.

Now you are ready to use your adapter. You should be able to slide the entire assembly over the microscope, and tighten the plastic thumbscrews to hold it in place.

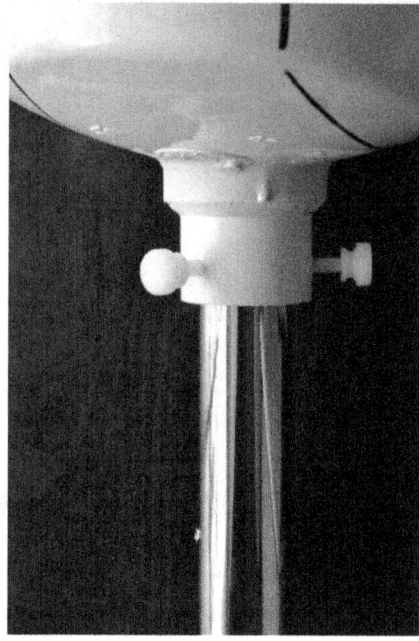

Tighten them evenly, by alternating one at a time, and do not over tighten.

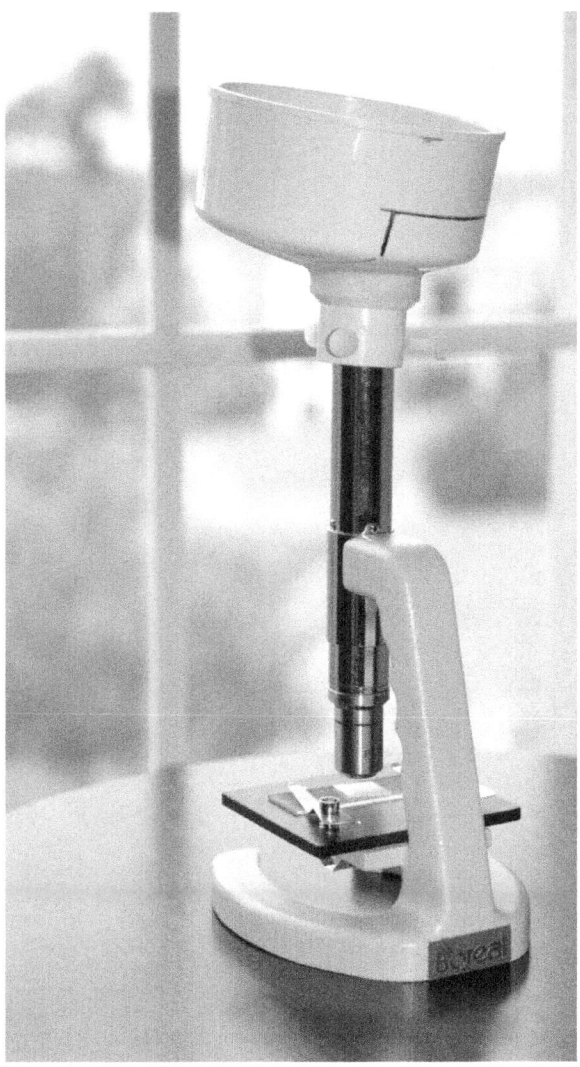

Turn on your camera, put it on macro setting, and drop it into place. All should work perfectly. Many of the microscope pictures in this book were taken using this same PVC adapter you see here.

CHAPTER 11

How to Make a Plant Press

In my research on tardigrades, it was important to describe and document the environment where tardigrades are found. Since most of my field samples were of lichen on trees, tree identification became all important. Field manuals about trees are very helpful, but often you will need to do more research to identify a particular tree species. Leaf samples are critical in this task. A good plant press is the answer.

Below you will find an outline of the very simple process of making a plant press. We'll start with materials.

Selecting Your Components

- Long bolts (4)
- Washers (8)
- Thumbscrews (4)
- Wood planks (2)

- Blotter paper sheets, acid free (1 pad)

Your wood planks must be wide and long enough to handle paper of a size at least 9 inches by 12 inches, plus the planks must allow for the bolts in each corner. Thus, your planks should be at least 14 inches by 18 inches, or slightly more than that.

Pick out two nice planks of wood, and keep in mind, you'll probably have this press for the rest of your days, and maybe pass it on to someone. So get some nice wood that you can sand, stain, and varnish. My daughter actually decorated my press using wood burning technique, which added to the beauty of the natural grain and dark knots in the wood.

With the exception of the wood planks, there is no need to measure very much. Just go to the hardware store, and look at their bolts, slide on the washers, and screw on the thumbscrews. The diameter of the long bolts does not matter, as long as the washers fit nicely and the thumbscrews work on the ends. The length of the bolts will depend upon how thick you want to make your press. I used very long bolts because I built a thick press. I used heavy scientific journals for top and bottom padding. Total cost, excluding wood, should be under $10.00.

For your actual pressings, buy acid-free paper in an art supply store. I used a thick pad of 9 inch by 12 inch archival quality watercolor paper.

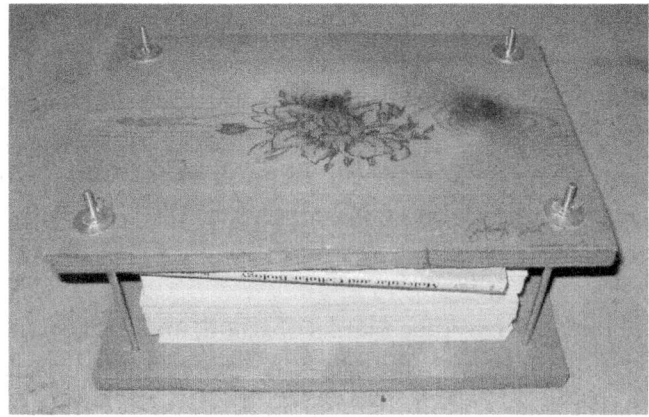

Building Your Plant Press

Drill four holes, one in each corner of your press. Be sure to space the holes far enough apart so that your paper, when inserted, will be clear of the holes. Stack your two planks to do the drilling so the holes are aligned perfectly. Then mark the edges of the planks so you will be able to realign the two planks the same way later when the paper is in-between.

Under the bottom plank, place the four washers at the hole positions. Then slide the bolts, from bottom up, through the washer and then the wood.

Lay the top plank on top of the four bolts, sliding it down flush with the bottom plank.

Drop the remaining four washers over the bolts, and screw on the thumbscrews, about 1/2 inch.

Your Microscope Hobby

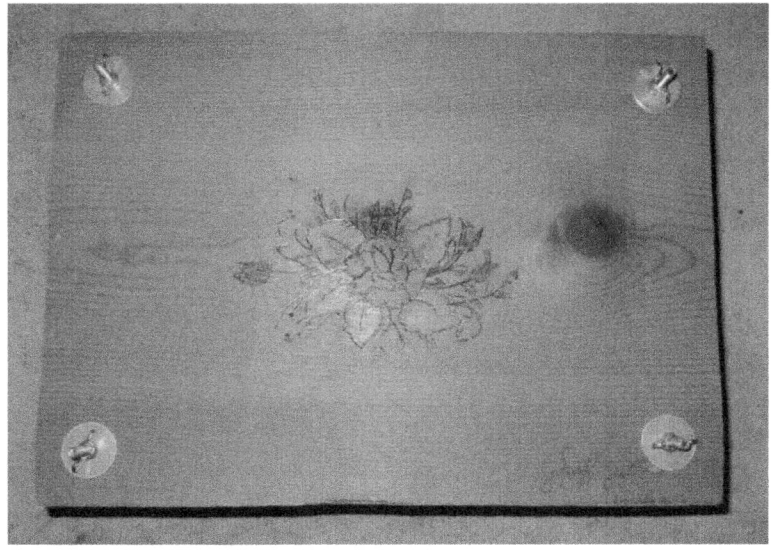

Now raise the top plank all the way up, as you insert your paper. You only need the archival paper in the center, and you can use old magazines and scrap sheets for top and bottom pressure.

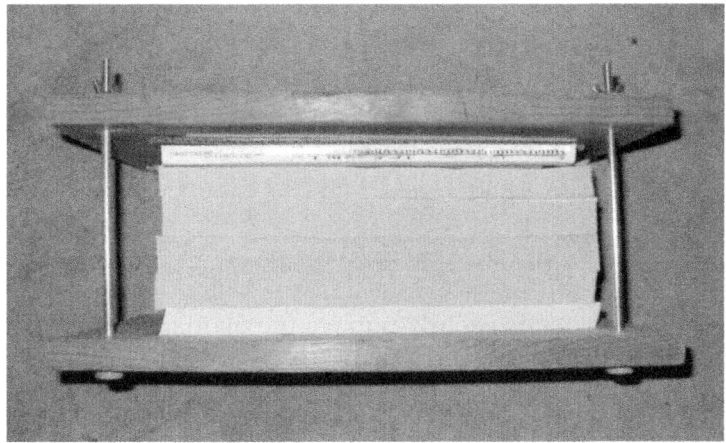

Now, enjoy taking flower, leaf and plant samples. After pressing for a couple of days, you can assemble the pressings in a book. You can even make color photocopies which come out beautifully, then scan those photocopies into .pdf files for research or posterity.

Here is another picture of a pressing.

You can also gently remove the plants, and preserve them using clear hot-lamination material. This will allow you to study both sides of the plant, and even put it under a microscope (if you slice up the laminate into small microscope slide-sized pieces).

Websites

Get your FREE color version of this book in electronic format. If you purchased this paperback though Amazon, then you will automatically be able to download the color Kindle version for free. If you would like a color version for another type of e-reader in mobi or epub format, or if you want a color pdf version to read on your computer desktop, here is the link:

http://www.mikeshawtoday.com/freebook/

If you have any difficulty downloading from the link, simply send the author an email, and request the version you would like. Send your request to:

freebook@mikeshawtoday.com

Here are all the other websites mentioned in this book, in order of appearance without duplicating:

Microscope store on Amazon

http://astore.amazon.com/mikesmicroscopestore-20

Nikon's annual photomicrography contest

http://www.nikonsmallworld.com/

Olympus' annual photomicrography contest

http://www.olympusbioscapes.com/

Pizza Express in England (used micro-photograph décor)

http://www.tripadvisor.com/Restaurant_Review-g504170-d732740-Reviews-Pizza_Express_Epsom-Epsom_Surrey_England.html

GoDaddy website domains and hosting

http://www.godaddy.com/default.aspx?isc=IAPtno100

Microscopy E-Magazine in England

http://www.microscopy-uk.org.uk/

Microscopy E-Magazine in England (article showing Rheinberg)

http://www.microscopy-uk.org.uk/mag/indexmag.html?http://www.microscopy-uk.org.uk/mag/artjan05/jmcbry02.html

Microscopy E-Magazine in England (article showing Oblique Illumination)

http://www.microscopy-uk.org.uk/mag/indexmag.html?http://www.microscopy-uk.org.uk/mag/artfeb00/pjoblique.html

Microscopy E-Magazine in England (article showing various types of illumination)

http://www.microscopy-uk.org.uk/mag/indexmag.html?http://www.microscopy-uk.org.uk/mag/artdec03/wdonion2.html

Fine Art America

http://fineartamerica.com/

Video of First Animal to Survive in Space

http://tardigrade.us/2012/09/04/mike-shaws-interview-on-tardigrades-and-space/

TV GLOBO in Brazil Sunday evening show appearance

http://tardigrade.us/2015/01/04/space-bear-hunter-on-brasilian-tv-globo/

National High Magnetic Field Laboratory interactive tutorial on Rheinberg

http://micro.magnet.fsu.edu/primer/techniques/rheinberg.html

Definition of a cork borer

http://en.wikipedia.org/wiki/Cork_borer

Axminster Tools

http://www.axminster.co.uk/boehm-boehm-hollow-punch-sets-prod22429/

Mayhew Tools

http://www.mayhew.com/

Silhouette Portrait Machine

http://www.silhouetteamerica.com/

Grainger website

www.grainger.com

American Science and Surplus on-line catalog

www.sciplus.com

Utrecht Art supply website

http://www.utrechtart.com/index.cfm

Hardware Store on-line source for Con-Tact ® brand self-adhesive vinyl

http://www.hardwarestore.com/

ULINE Company website (source for tiny plastic bags)

www.uline.com

The Grafix Arts website

http://www.grafixarts.com/product/ClearLay

Lee Filters website

www.leefilters.com

Roscolux Filters website

http://www.rosco.com/us/index.cfm

Stage Spot website

http://www.stagespot.com/

Cheap Light website

http://www.cheaplights.com/cart/page31.html

Premier Lighting website

http://www.premier-lighting.com/sales/colorfltr.html

Adorama Camera website

www.adorama.com

Three Rivers Archery

http://www.3RiversArchery.com

Micscape Magazine Photo Gallery

http://www.microscopy-uk.org.uk/mag/indexmag.html?http://www.microscopy-uk.org.uk/mag/artjun05/swgallery2.html

Microscopy-UK Main Page

http://www.microscopy-uk.org.uk/

DIY/DIC article by Wim van Egmond

http://www.microscopy-uk.org.uk/mag/artnov02/diydic.html

Oblique Article in Micscape

http://www.microscopy-uk.org.uk/mag/indexmag.html?http://www.microscopy-uk.org.uk/mag/artfeb00/pjoblique.html

Rheinberg Filters Website

http://www.rheinberg-filters.com

Tardigrade website

www.tardigrade.us

Mike Shaw's website

www.mikeshawtoday.com

Yahoo Groups

http://groups.yahoo.com/search?query=microscope

Acknowledgements

Thank you to my wife Donna, a great partner in business and life, and my two daughters who keep it real.

BOOKS

Kids & Teachers Tardigrade Science Project Book
Kids & Teachers Tardigrade Quiz & Fact Book
WORD NERD Riddle Game Book About Things Way Up High

CONTACT INFORMATION

Mike Shaw
PO Box 742
Midlothian, VA 23113
www.tardigrade.us
www.mikeshawtoday.com

Your Microscope Hobby – How to Make Multicolored Filters ©, Copyright Michael W. Shaw, 2014

THE AUTHOR

The author holding a mallet made from a gum tree.
He uses this to punch out really big filters.

Made in the USA
Monee, IL
10 December 2020